Gender, Development and Social Change

Series Editor
Wendy Harcourt, The International Institute of Social Studies, Erasmus University, The Hague, The Netherlands

The *Gender, Development and Social Change* series brings together path-breaking writing from gender scholars and activist researchers who are engaged in development as a process of transformation and change. The series pinpoints where gender and development analysis and practice are creating major 'change moments'. Multidisciplinary in scope, it features some of the most important and innovative gender perspectives on development knowledge, policy and social change. The distinctive feature of the series is its dual nature: to publish both scholarly research on key issues informing the gender and development agenda as well as featuring young scholars and activists' accounts of how gender analysis and practice is shaping political and social development processes. The authors aim to capture innovative thinking on a range of hot spot gender and development debates from women's lives on the margins to high level global politics. Each book pivots around a key 'social change' moment or process conceptually envisaged from an intersectional, gender and rights based approach to development.

Wendy Harcourt · Ana Agostino ·
Rebecca Elmhirst · Marlene Gómez ·
Panagiota Kotsila
Editors

Contours of Feminist Political Ecology

palgrave
macmillan

Editors
Wendy Harcourt
International Institute of Social Studies of Erasmus
University Rotterdam
The Hague, The Netherlands

Rebecca Elmhirst
University of Brighton
Brighton, UK

Panagiota Kotsila
Universitat Autònoma de Barcelona
Barcelona, Spain

Ana Agostino
University CLAEH
Montevideo, Uruguay

Marlene Gómez
Freie Universität Berlin
Berlin, Germany

This project received funding from the European Union's Horizon 2020 research and innovation programme under the Marie Sklodowska-Curie grant agreement No. 764908-WEGO 2018-2021

ISSN 2730-7328 ISSN 2730-7336 (electronic)
Gender, Development and Social Change
ISBN 978-3-031-20927-7 ISBN 978-3-031-20928-4 (eBook)
https://doi.org/10.1007/978-3-031-20928-4

Cover credit: WOMEN SPACES INVIO

This Palgrave Macmillan imprint is published by the registered company Springer Nature Switzerland AG
The registered company address is: Gewerbestrasse 11, 6330 Cham, Switzerland

Foreword

We are living in troubling times: the ongoing coronavirus pandemic and its resulting massive loss of lives and livelihoods, combined with ongoing climate, health, water and related crises. Feminist political ecology (FPE) traditionally provides powerful tools that show how environmental "crises", environmental change and human–environment relations are determined by power, gender, class, race, ethnicity, caste and specific socio-economic, cultural and historical legacies (Elmhirst, 2011; Harcourt & Nelson, 2015; Rocheleau et al., 1996). With a focus on the wider political-economic context and both micro to macro scales, FPE has called out injustices due to processes of allocation, dispossession and extractivism, unequal power and gender relations and the uneven spread of pains and gains from development processes (Elmhirst, 2018). A focus on cognitive and epistemic justice has allowed FPE to lift invisibilised and vulnerable voices and perspectives from the margins; voices that are not acknowledged in dominant policy processes and mainstream knowledge production (Nightingale, 2013). This book builds on FPE's capacity to "see multiple" and seek out diverse knowledges (Rocheleau, 2015) as it continues to elaborate on the gendered nature of environmental protest and activism, imagine alternative visions and engage with feminist critiques of mainstream science and epistemology (Leach et al., 2015; Rocheleau, 2015).

The book was conceptualised and written by members of the Well-being, Ecology, Gender and cOmmunity Innovative Training Network

(WEGO), a group which provided the foundation for rich intergenerational and situated conversations that help make sense of our troubled times. I consider myself lucky and honoured to have been part of this network of scholar activists that focused not only on doctoral training, but also on action research and advocacy around communities, commons and commoning. I learnt deeply from the often intense interactions, discussions and meetings, both in person and online. By taking a multi-faceted view of FPE, this book captures the diverse standpoints and perspectives contained in the WEGO network, ranging from more traditional political ecology issues such as extractivism, land and water to more recent engagements around embodied experiences of pain and ageing, the ethics of more-than-human interdependencies, indigenous ontologies of co-becoming and feminist perspectives around degrowth, decoloniality and pluriversality. The book is rooted in the feminist tradition of praxis, reflexivity and a commitment to gender, social and environmental justice. It foregrounds the importance of care for human, more-than-humans and the planet, not least because the extraordinary loss and disruptions during the ongoing pandemic forced researchers to become aware of the need to care for each other, often through digital means (Di Chiro, 2017). FPE, as conceptualised in the book's conversations, not only explores the ethics and politics of care in research, activism and everyday life but also teases out the often-overlooked non-material aspects of political ecology and degrowth (care, love, solidarity, reciprocity) that allow for human and environmental well-being to flourish (Mehta & Harcourt, 2021). Many of the chapters focus on aspects of care, living and being that both reset the growth imaginary and inform radical change, contributing to visions of societies based on caring relations, well-being and reciprocity. The book also asks uncomfortable questions regarding unequal power and privilege, for example the difficult question of who needs to and can afford to degrow when speaking about the degrowth movement (Barca, 2020).

Despite the challenges of the pandemic, which led to the authors moving to digital research for two years, the different chapters provide engaging dialogues and provoking conversations across different continents, standpoints and contexts. The authors give powerful responses to authoritarianism, racism, patriarchy, extractivism, climate crises, disease and ageing, often in creative ways: for example, through co-curating exhibitions, using artistic and visual approaches and storytelling around land, water, food, energy and bodies. They link experiences of climate change with processes of extraction, exploitation and resistance in communities in the global South and North, based on their situated knowledges and lived experiences as citizens, climate and environmental activists and researchers. In this volume, authors deploy FPE to challenge the biomedical gaze and engage with the political ecologies of health, disease and environmental politics through situated accounts of living with chronic pain, illness, disease and environmental injustices. The dialogues across geographies, generations and disciplines draw on personal intellectual journeys, histories and experiences and provide grounded insights on diverse topics. These range from views of decoloniality, to a feminist perspective on the controversies around global population, to inspiring visions for non-hegemonic alternative urban energy futures, to non-western (Andean) cosmovisions on ways of living, being and the relationships between the social and natural world.

Together the chapters push the FPE project of foregrounding multiples ways of knowing and being, thus enabling new conceptions of politics, justice and alternatives to dominant, capitalist development trajectories. The book thus draws on FPE scholarship and tools to deal with contradictory and difficult questions, thinking also about how transformative research can help create more equitable and just worlds. Through its diverse dialogues and conversations, this remarkable book is a great contribution to the collective anti-capitalist feminist struggle against planetary destruction and for a pluriverse that will allow beings to flourish and thrive.

As Ojeda et al. (2022) state: "The powers arrayed against ecological and social justice are formidable in their magnitude and entrenchment ... In this struggle against inequalities, injustices, degradations, exploitations, and planetary destruction, we all need to be on the same side despite our differences and partial knowledges" (Ojeda et al., 2022, pg. 17).

Contours of Feminist Political Ecology shows not only the depth of those challenges but just how important it is that a passionately engaged, feminist (academic) stance can build the foundation for conversations that enable us to listen and learn from each other and our differences.

Brighton, UK Lyla Mehta

References

Barca, S. (2020). *Forces of reproduction: Notes for a counter-hegemonic Anthropocene*. Cambridge University Press.

Di Chiro, G. (2017). Welcome to the White (M)anthropocene?: A Feminist-environmentalist Critique. In S. MacGregror (Ed.), *Routledge handbook of gender and environment* (pp. 487–505). Routledge.

Elmhirst, R. (2011). Introducing new feminist political ecologies. *Geoforum, 42*(2), 129–132.

Elmhirst, R. (2018). Ecología política feminista: perspectivas situadas, compromisos emergentes. *Ecologia Politica* 54. Retrieved from https://www.ecologiapolitica.info/ecologias-politicas-feministas-perspectivas-situadas-y-abordajes-emergentes/

Harcourt, W., & Nelson, I. L. (Eds.). (2015). *Practicing feminist political ecology: Moving beyond the green economy*. Zed Books.

Leach, M., Mehta, L., & Prabhakaran, P. (2015). Sustainable development: A gendered pathways approach. In M. Leach (Ed.), *Gender equality and sustainable development* (pp. 19--51). Routledge.

Mehta, L., & Harcourt, W. (2021). Beyond limits and scarcity: Feminist and decolonial contributions to degrowth. Political Geography, 89, 102411. https://doi.org/10.1016/j.polgeo.2021.102411

Nightingale, A. (2013). A feminist in the forest: Situated knowledges and mixing methods in natural resource management. *ACME: An International Journal for Critical Geographies, 2*(1), 77–90.

Ojeda, D., Nirmal, P., Rocheleau, D., & Emel, J. (2022). Feminist Ecologies. *Annual Review of Environment and Resources* 47. https://doi.org/10.1146/annurev-environ-112320-092246

Rocheleau, D. (2015). A situated view of feminist political ecology from my networks, roots and territories. In W. Harcourt & I. L. Nelson. (Eds.), *Practising Feminist Political Ecologies: Moving Beyond the Green Economy* (pp. 29–66). Zed Books.

Rocheleau, D., Thomas-Slayter, B., & Wangari, E. (1996). A feminist political ecology perspective. In Rocheleau, D., B. Thomas-Slayter, & E. Wangari (Eds.), *Feminist political ecology: Global issues and local experiences* (pp. 3–26) Routledge.

Lyla Mehta is a Professorial Fellow at the Institute of Development Studies of the University of Sussex, UK and an Adjunct Professor at the Department of International Environment and Development Studies (Noragric), Norwegian University of Life Sciences. She is a mentor and supervisor in the WEGO Innovative Training Network. Her work focuses on water and sanitation, forced displacement and resistance, climate change, scarcity, rights and access, resource grabbing and the politics of environment/development and sustainability.

Acknowledgements

We would like to thank all the allies, friends, partners, administrators, mentors and communities who participated in shaping this book, and the whole Wellbeing Ecology Gender cOmmunity (WEGO)-ITN experience. A very special thank you to Sharmini Bisessar-Selvarajah who together with others at the International Institute of Social Studies of Erasmus University Rotterdam and WEGO-ITN partner institutions worked hard to ensure the running of the project through all its trials and tribulations, especially during the COVID-19 Pandemic.

This project received funding from the European Union's Horizon 2020 research and innovation programme under the Marie Sklodowska-Curie grant agreement No. 764908-WEGO 2018-2021.

About This Book

This edited book sets out the contours of feminist political ecology that have emerged from the multi-sited, cross-generational and inter/transdisciplinary dialogues held in the Wellbeing Ecology Gender cOmmunities (WEGO) network over the last four years. It sets out to show how feminist political ecology is a vibrant and engaged research process at the intersection of feminist and environmental theory and practice.

It covers topics that range from climate change and extractivism, to body politics and health, degrowth, care and community well-being. The book maps out the contours of feminist political ecology based on the WEGO network's original research and analysis as it links conversations with local communities, in social movements as well as within different academic spaces.

The book illustrates how FPE scholarship is shaped by everyday and embodied lives within damaged, dynamic and contested environments. Key to the book is how FPE can produce feminist intersectional and intergenerational plural knowledges as a political and ethical practice, recognising multiple truths, intergenerational and intersectional differences. The book is made up of a series of dialogues that navigate theory and practice, individual and collective engagements at the edge of academic and activist desires to produce politically meaningful knowledge building on the insights of empirical research and feminist theory.

Contents

Notes on Contributors

Ana Agostino (co-editor) has a Ph.D. in Development Studies from the University of South Africa (UNISA) and is ombudsperson and mentor of the WEGO Innovative Training Network. She is a Lecturer at University CLAEH (Uruguay) in Development, Culture and Human Rights and at FLACSO (Latin American Faculty of Social Sciences) in Gender and Environment. In 2022 she was a Guest Professor and Mercator Fellow at the University of Kassel, Germany and since 2017 she is the Vice-president of the Latin American Institute of Ombudsman Institutions.

Stefania Barca is Distinguished Researcher at the University of Santiago de Compostela (Spain) and a mentor and partner in the WEGO Innovative Training Network. An environmental historian and feminist political ecologist, originally from Italy, she investigates the labour/gender/ecology nexus in the industrial age and the politics of Just Transition. In 2021, she held the Zennström Professorship in Climate Change Leadership at Uppsala University, in Sweden. She is active in the degrowth feminist network (FADA) and in the international Care Income campaign.

Nick Bourguignon is a Ph.D. student of the WEGO Innovative Training Network at the Institute of Environmental Science and Technology at the Autonomous University of Barcelona (ICTA-UAB). He graduated from the University of Warwick in the UK and the International Institute of Social Studies of Erasmus University Rotterdam in The Netherlands. His current research explores the socionatural transformations of irrigation territories and communities linked by inter-basin water transfers in Spain.

Giovanna Di Chiro is Professor of Environmental Studies at Swarthmore College where she teaches courses on environmental justice theory, ecofeminism(s), action research methods, and community sustainability. She is a mentor and partner of WEGO-ITN and co-founder and faculty partner with Serenity Soular, a campus–community collaborative in North Philadelphia. She is currently working on her next book *Worldmaking from the Ground Up: The Praxis of Environmental Justice.*

Constance Dupuis is a Ph.D. student of the WEGO Innovative Training Network feminist researcher at the International Institute of Social Studies of the Erasmus University Rotterdam where she also did her Masters. She is currently doing a Post-Doc at McMasters University, Canada. Her research focuses on ageing, well-being in later life and intergenerational care.

Dian Ekowati is a Ph.D. student of the WEGO Innovative Training Network at the University of Brighton. She researches everyday care discourse in the extractivist landscape titled: Gendered Analysis of Care in Family Farms in Oil Palm Policy Discourse in Indonesia. Dian believes in the in-between spaces of her Ph.D. research and uses autoethnography (among other methods) to approach care in her research. Prior to joining WEGO, she worked to support research and community projects with CIFOR, the Sajogyo Institute and the Samdhana Institute in Indonesia.

Rebecca Elmhirst (co-editor) is Professor of Human Geography in the Centre for Spatial, Environmental and Cultural Politics at the University of Brighton in the UK, where she teaches undergraduate courses on feminist political ecology, sustainability and climate justice. She is a mentor and supervisor of the WEGO Innovative Training Network and is co-editor of the journal Gender, Technology and Development. Her research focuses on everyday politics, land and livelihoods in contested environments, and she works through partnerships with colleagues in Indonesia and Thailand.

Khayaat Fakier is Senior Lecturer in Sociology at Stellenbosch University and Prince Claus Chair in Equity and Development at the International Institute of Social Studies of Erasmus University Rotterdam in The Hague from 2021–2023. Her research interests are on care, social reproduction, gender feminisms and migration.

Eoin Farrelly is a Ph.D. student of the WEGO Innovative Training Network at the Department of Sociology & Human Geography, University of Oslo. His research examines political agonisms in initiating socio-ecological transformations in fishing communities in Scotland. His academic background is in the study of peace, violence and gender. He is on the boards of BeLonG To Youth Services, Ireland's leading LGBTQ+ youth service and the Irish Family Planning Association.

Mila Fenner is a Ph.D. student of the WEGO Innovative Training Network at the International Institute of Social Studies (ISS) of Erasmus University Rotterdam in The Hague. She graduated from the University of Oxford and the University of Cambridge. After graduating she worked in research communication for the Food and Climate Research Network in Oxford and TRAFFIC, an international NGO focusing on curbing illegal trade in wild animals and plants. She has conducted field work in India, Malaysia and Cameroon and is a part-time director and writer for film and theatre.

Marlene Gómez (co-editor) is a Ph.D. student of the WEGO Innovative Training Network at the division of Gender and Diversity of the Freie Universität Berlin. Her research is on feminist political ecology and the politics of food in urban southern Europe.

Wendy Harcourt (co-editor) is Professor of Gender, Diversity and Sustainable Development at the International Institute of Social Studies (ISS) of Erasmus University Rotterdam in The Hague. She is coordinator, mentor and supervisor of the WEGO Innovative Training Network. She joined ISS in November 2011 after 23 years as editor and director of programmes at the Society for International Development, Rome, Italy. She has published a prize wining monograph, 13 edited books, and written widely in critical development studies, gender and development and feminist political ecology.

Elona M. Hoover is an independent researcher with a doctorate in human geography from the University of Brighton. Her work weaves together research, activism and poetic practices, most recently on issues of urban commoning, emancipatory affective practices, relational ethics, community-grounded food and agriculture and the feminist political ecologies of extractivism.

Ilenia Iengo is a Ph.D. student of the WEGO Innovative Training Network at the Barcelona Laboratory for Urban Environmental Justice and Sustainability, ICTA-UAB. Her action research is situated in Naples, Italy with a focus on emancipatory urban politics and imaginaries sprouting at the intersection of transfeminism and environmental justice. Her work is published on *Environmental Humanities, Journal of Political Ecology and Environmental Justice.* She is a member of Undisciplined Environments and Ecologie Politiche del Presente collectives.

Panagiota Kotsila (co-editor) is a Juan de la Cierva-incorporacion fellow at ICTA-UAB where she is leading research on urban feminist political ecology. She is a mentor and supervisor in the WEGO Innovative Training Network and a core researcher at the Barcelona Lab for Urban Environmental Justice and Sustainability (BCNUEJ), a member of the Undisciplined Environments collective.

Seema Kulkarni is a Mentor of the WEGO Innovative Training Network and holds an MA degree in Social Work from Tata Institute of Social Sciences, Mumbai. She is one of the founding members of Society for Promoting Participative Eco-system Management, Pune (SOPPECOM). She has co-ordinated various studies and programmes around decentralisation, gender and land, water and sanitation.

Irene Leonardelli is a Ph.D. student of WEGO Innovative Training Network at the department of Water Governance at IHE Delft Institute for Water Education, in Delft, The Netherlands. Before starting her Ph.D., she worked on several research projects related to migration dynamics across the Mediterranean Sea and environmental migration. Her current research examines processes of rural agrarian transformation and water governance in Maharashtra, India, from a feminist perspective.

Siti Maimunah was awarded her Ph.D. in December 2022 from the University of Passau, Germany as a member of the WEGO Innovative Training Network. She is an activist and researcher working alongside NGOs to support communities affected by extractivism projects in Indonesia with JATAM, TKPT, and Sajogyo Institute. Her research combines activism and academic works to understand how the configuration of the resource frontier through mega projects such as mining and logging shaped ethnicity, gender, and intersectionality on Kalimantan Island, Indonesia.

Lyla Mehta is a Professorial Fellow at the Institute of Development Studies of the University of Sussex, UK and an Adjunct Professor at the Department of International Environment and Development Studies (Noragric), Norwegian University of Life Sciences. She is a mentor and supervisor in the WEGO Innovative Training Network. Her work focuses on water and sanitation, forced displacement and resistance, climate change, scarcity, rights and access, resource grabbing and the politics of environment/ development and sustainability.

Iliana Monterroso holds a Ph.D. in Environmental Sciences. She is a mentor and partner in the WEGO Innovative Training Network and a consultant for the Climate and Land Use Alliance, CLUA, leading a program on resilience in Mexico and Central America. Her work focuses on political ecology, tenure, justice and rights specially in Latin America.

Nanako Nakamura is a Ph.D. student of the WEGO Innovative Training Network at the Department of Social Sciences, Wageningen University, The Netherlands. Her research focuses on care, well-being and post-capitalist community economies, at the intersections of age, rurality and gender in Japan.

Ingrid L. Nelson is an Associate Professor in the Department of Geography and Geosciences at the University of Vermont. Her work on feminist political ecology is published in *Gender, Place & Culture*, *Environment and Planning E: Society and Space*, *Geoforum*, and other journals and collections. She is a mentor and partner in the WEGO Innovative Training Network and is currently collaborating on several projects regarding feminist digital natures.

Andrea J. Nightingale is Professor of Human Geography, University of Oslo and Research Fellow, Swedish University of Agricultural Sciences and a mentor and supervisor in the WEGO Innovative Training Network. Her current research looks at power and politics within dynamic and unpredictable environmental change. Her interests include climate change adaptation and transformation debates; collective action and state formation; the nature–society nexus; political violence in natural resource governance; and feminist work on emotion and subjectivity in relation to development, transformation, collective action and cooperation.

Mariana Jesús Ortecho is a Senior Researcher at the Argentinian National Council for Scientific and Technical Research and an Associate Professor at the Chair "Problems of Theatre and Culture" and "Culture and Society in Latin America" in the school of Arts of the National University of Cordoba, Argentina where she completed her Ph.D. Her research focuses on the language-knowledge relationships in the practices of science and art within the framework of Western civilization.

Alice Owen is a Ph.D. student of the WEGO Innovation Training Network at the University of Brighton. Alice looks at the knowledge politics of fracking and unconventional fossil fuel extraction in the UK, critically examining whose knowledge comes to matter in this unfolding conflict over the environment and over contemporary democracy.

Chizu Sato teaches in the Cultural Geography group at Wageningen University in The Netherlands. She is a mentor and supervisor in the WEGO Innovative Training Network and was on the coordination team of the 8th International Degrowth Conference: Caring Communities for Racial Change. She is co-editor of *Rethinking Marxism: A Journal of Economics, Culture, and Society*. Her research interests include transnational feminism, Marxism, international development and political ecology, where she focuses on the intersections of women, empowerment and development.

Ankita Shrestha is a Marie Skłodowska-Curie Ph.D. Research Fellow at the University of Oslo. Her PhD thesis presents alternative views on the "political subject".

Lillian Sol Cueva is a Ph.D. researcher at the International Institute of Social Studies-Erasmus University Rotterdam, The Netherlands. Her research uses storytelling and art-based methods to study energy futures from a feminist perspective. Previous experience includes advocacy in human and women's rights, climate change, mobility, urban development and energy.

Agustina Solera is a Postdoctoral Researcher of the Prince Claus Chair in equity and development and lecturer at the International Institute of Social Studies of Erasmus University Rotterdam in The Hague. She completed her Ph.D. in Social Studies of Latin America in the National University of Córdoba, Argentina. Her work bridges the critical studies of habitat and the broad transdisciplinary field of Latin American critical thinking.

Enid Still is a Ph.D. student of the WEGO Innovative Training Network at the University of Passau in Germany. Before starting her Ph.D., she studied anthropology at the University of Oxford, UK and Savitribai Phule Pune University, India, where her research focused on the politics of hope in theatre and development practice. Her current research employs a feminist and multidisciplinary approach to explore the entanglements of colonialism, agriculture and ethics, specifically their histories and presents in Tamil Nadu, India.

Anna Katharina Voss is a Ph.D. student of the WEGO Innovative Training Network at the International Institute of Social Studies of Erasmus University Rotterdam. She has participated in volunteering and research projects in the field of alternative agriculture, community activism and permaculture in Europe, Africa and India. Currently she is looking at the entanglements of agroecology and feminism in rural–urban grassroots organisations and rooted networks through an FPE lens.

Eunice Wangari Muneri is a Ph.D. student of the WEGO Innovative Training Network at the Institute of Development Studies at the University of Sussex, UK. Eunice has a Master's from Wageningen University and from Jomo Kenyatta University of Agriculture and Technology, Kenya. Eunice has worked as a micro-credit officer at a local bank in Kenya and trained women farmers in implementing climate-smart agriculture initiatives at a grassroots community-based organisation in Kenya.

Abbreviations

DT	Decolonial Thinking
FPE	Feminist Political Ecology
WEGO-ITN	Wellbeing, Ecology, Gender and cOmmunity Innovative Training Network

List of Figures

CHAPTER 1

Sketching Out the Contours

Ana Agostino, Rebecca Elmhirst, Marlene Gómez, Wendy Harcourt, and Panagiota Kotsila

Introduction

This book is a contribution to the vital response of feminist political ecology (FPE) to global environmental, climate, health, economic and political crises and their impact on life in all its diversity. We sketch out how FPE is responding to these crises based on a series of multi-sited, cross-generational and inter/transdisciplinary dialogues held over the last four years in the Wellbeing, Ecology, Gender and Community Innovative Training Network (WEGO-ITN). In the tradition of FPE, the book embraces a deliberately open-ended approach. It does not intend to define or delimit FPE through a series of descriptive analyses of key concepts or specific case studies, or a showcasing of new methodologies. Instead, it is

A. Agostino
University CLAEH, Montevideo, Uruguay
e-mail: anaa@internet.com.uy

R. Elmhirst
University of Brighton, Brighton, UK
e-mail: r.j.elmhirst@brighton.ac.uk

W. Harcourt et al. (eds.), *Contours of Feminist Political Ecology*,
Gender, Development and Social Change,
https://doi.org/10.1007/978-3-031-20928-4_1

a series of conversations that take up the concerns and troubles that have emerged around the WEGO-ITN collaborative process.

The book is made up of a series of dialogues that navigate theory and practice, individual and collective engagements at the edge of academic and activist desires to produce politically meaningful knowledge, building on the insights of empirical research and feminist theory. The chapters cover on-line exhibitions on extractivism, intimate discussions on embodied experiences of ill health and pregnancy, discussions around ageing and care, storytelling and emotional engagements with water and contested academic debates around population and political protests. The variety of situations, places and concerns illustrates in numerous ways in which FPE scholarship is shaped by everyday and embodied lives within damaged, dynamic and contested environments, as well as by hope in collaborative ventures on the margins of academic practices. The book shows how FPE as a convening space fostering vibrant and engaged research processes at the intersection of feminist and environmental theory and practice.

Acknowledging the importance of these embodied experiences, FPE promotes grounded and engaged research to understand and make visible political processes, including the emotions and embodied reactions and responses of people and communities to economic, social and environmental change. It recognises that knowledge is generated on different scales, including day-to-day experiences. FPE is in dialogue with various feminist and environmental justice communities, as well as the academic

M. Gómez
Freie Universität Berlin, Berlin, Germany
e-mail: marlene.gomez@fu-berlin.de

W. Harcourt (✉)
International Institute of Social Studies, Erasmus
University Rotterdam, The Hague, The Netherlands
e-mail: harcourt@iss.nl

P. Kotsila
Institute of Environmental Science and Technology (ICTA), Universitat
Autònoma de Barcelona (UAB), Barcelona, Spain
e-mail: panagiota.kotsila@uab.cat

community, and strives to do non-extractive research through a participatory process of co-production which translates multiple experiences with the environment with respect and care. FPE challenges us to ask: what are the ethics and moral norms that shape our knowledge production? How can we use that knowledge to build culturally safe and secure spaces that recognise situated struggles, multiple imaginaries and bodily diversities? The book itself reflects this with different takes on what FPE means in both theory and practice, hence the authors are intentionally explicit about the context, embodied and epistemic positioning of their different voices and experiences.

In this introduction, we first position the book's discussions with other FPE texts. We then look at how the book evolved from the WEGO-ITN experience of open-ended intergenerational and inter/transdisciplinary conversations, followed by a summary of the key themes that emerge in the book chapters. In conclusion, we ask, 'where do we go from here?' and indicate some future directions for FPE research and practice.

Engaging with FPE Conversations

The conversations in Feminist Political Ecology (FPE) with which this book engages are as broad as they are deep. We find Diane Rocheleau's metaphor of 'rooted networks' (Rocheleau, 2015) useful for capturing the breadth and range of our engagements in this diverse and open-ended area of practice and thought. The debates that emerge around attempts to chart and define FPE are testament to the situatedness of knowledge claims that are made when mapping such conversations. For example, Elmhirst's (2011) review began from Rocheleau et al.'s (1996) landmark text that first coined the phrase Feminist Political Ecology, and from this starting point, Elmhirst's review centred on works that brought feminist theorisations of subjectivity to bear on issues of dispossession, resource access and control. Taking a similar tack, but more mindful of FPE's relationship with global South feminisms, Resurrección (2017) considers FPE through the relationship between gender and development studies and activist ecofeminisms, while Harcourt and Nelson (2015) narrate FPE through queer ontologies and post-humanist body politics. Other scholars have pressed further to challenge the hegemonies of white, Anglophone knowledge practices to tell a story of FPE through environmental justice and critical race theory (Mollett, 2017;

Mollett et al., 2020) and through decolonial and anticolonial ecological feminist thinking and practice (Ojeda et al., 2022; Sultana, 2021; Sundberg, 2017). Such reviews reflect the temporality of broader intellectual currents and political demands, the often hemispheric spatialities of authors' rooted networks and journeys between activism, academia and policy arenas. Stories *about* FPE reflect differently situated conversations *within* FPE.

Thus, in this section which positions and honours the histories and many conversations on which this book is built rather than presenting a singular narrative of Feminist Political Ecology's 'origins' and engagements, we introduce a multiplicity of starting points and ongoing conversations that have inspired the chapters in this book, and indeed, the WEGO project. These plural starting points reflect and continue to influence an ever emergent but enduring set of practices, values and ethical principles underpinning FPE, and provide a source for the concepts that guide and are developed in the academic practice of FPE. Mindful of the tyranny of exclusion and inclusion in any effort to 'map' FPE in the knowledges we value in the practice of academic citation, we reference the different texts here in order to point readers to the diverse conversations in which we engaged with FPE and, as Max Liboiron writes to indicate to readers 'the relations we want to build...enacting good relations in a text, through a text' (Liboiron, 2021, p. 1).

For some of us, FPE emerged out of Political Ecology and critical development studies, which focus on how power operates within socio-ecological relations and research practice. Here, inspiration has come from ecofeminism (Seager, 1993; Shiva, 1988), feminist science and technology studies (Harding, 1991) and postcolonial feminist critiques of development (Agarwal, 1998; Mohanty, 2003). FPE has engaged the Black feminist concept of intersectionality (Cho et al., 2013) to move beyond a singular and homogenising focus on women and/or gender binaries. Instead, attention is drawn to multiple *intersecting inequalities* (e.g., race, gender, class, ethnicity, religion, sexuality, age and geographical location) and how these are reproduced and perpetuated within particular landscapes (Mollett, 2017; Rocheleau et al., 1996). This has deepened analysis within 'traditional' political ecology concerns such as dispossession and extractivism (Mollett, 2018), and conservation, access and exclusion (Sundberg, 2004). Researchers within FPE have taken an intersectional analysis to explore ecological relations, developing conceptualisations that show how environmental subjectivities are forged through relations with nature and everyday material practices (Nightingale, 2011).

These conversations have embraced a focus on overlooked spheres of political ecological life, including lived experiences in the everyday (Harris, 2015) and embodied practices (Sultana, 2011) as sites of emotion, meaning and affect (Gururani, 2002; Singh, 2013). Space has opened to understand environmental conflicts beyond cognitive valuations of nature, where 'emotional political ecology' focuses attention on 'what one is allowed to remember, feel, enjoy or live' (González-Hidalgo & Zografos, 2020, p. 236; Velicu, 2015). Cross-cutting these themes, this branch of FPE has found inspiration from feminist critiques of science around who counts as a knowledge producer, what counts as knowledge and how knowledge is produced (Sundberg, 2017). FPE draws on Haraway's (2001) concept of situated knowledge, which unmasks the aura of scientific objectivity, and instead foregrounds the embodied social relations from which knowledge emerges (Nightingale, 2003). This conversation has been central in building FPE research practice that aims to be ethical and accountable, and to avoid reproducing power relations.

A third and related set of conversations in FPE emerges around questions of sufficiency, degrowth and commoning as launch points for reversing the damaging logics of unrelenting capitalist economic growth and extractive nature-society relations, understood as a crisis of social reproduction (Barca, 2020; Federici, 2004; Fraser, 2016). Inspiration here is drawn from feminist activism and scholarship that seeks instead to value all life-sustaining labour, thus foregrounding relations and economies of care (Tronto, 2015) and replacing efficiency with sufficiency (Wichterich, 2015). FPE draws on ecofeminist theoretical roots (Mellor, 2006; Shiva, 1988) to advocate ethical relations with more-than-human species and the natural world as part of a wider ethics of care (Di Chiro, 2017). Deeper engagements with narratives of restorative and transformative justice (Agostino, 2018) are explored, with the aim of living and being 'otherwise': fostering reciprocity, community and care (Puig de la Bellacasa, 2017). In engaging with these concerns, FPE is enriched through its links with the community economies research led by JK Gibson-Graham et al. (2020), recent organising around feminist degrowth (Dengler & Seebacher, 2019) and with Indigenous and other worldviews that question the Eurocentric paradigm of growth, instead expressing ideas of living and flourishing within ecological principles (Kothari et al., 2014; Nirmal & Rocheleau, 2019).

Finally, an important set of conversations in FPE emerges through engagement with decolonial thought, which addresses the epistemic privilege, violence and authority held in Eurocentric ways of knowing, being and doing (Fanon, 2001; Quijano, 2000). Decolonial conversations around feminism in FPE have explored how Western (Eurocentric) attempts to categorise, control and manage nature also involve control over racialised and gendered bodies (Lugones, 2010), expressed by some decolonial feminists through the concept of *cuerpo-tierra-territorio* (body–land–territory). In capturing the dynamics of racism, capitalism and gender oppression that underpin the coloniality of dispossessions, extractivism and embodied experience, similar connections are drawn between body, land and territory through different registers in other Indigenous places (Nirmal & Rocheleau, 2019). In attending to the coloniality of gender and the historicization of the Eurocentric hierarchization of (white) humans above more-than-human nature, decolonial perspectives in FPE opens space for alternative and pluriversal possibilities of 'being in relation'. From communities that fight socio-ecological harms through thinking and being 'otherwise', the prefigurative everyday practices within anticolonial feminist ecological activisms model ethical relationships with other species and the natural world (Candraningram, 2018) and inspire the practice of FPE.

In the next section we set out how each chapter is engaging in these diverse, and overlapping, conversations, but before we turn to that mapping out of the FPE threads weaving through the book, we need to explain how the book emerged through the WEGO-ITN which many of the authors belong.

WEGO-ITN: Collective Learning Process

In this section we take a moment to reflect on how WEGO-ITN's collective learning process helped to map out new areas and methodological approaches in FPE as set out in the book. WEGO-ITN ran from 2018 to 2022, funded by the EU. With 15 PhD projects at its core, it evolved into a network of scholar-activists working on feminist political ecology from institutions in Germany, Italy, Norway, Spain, The Netherlands, the UK in the EU and outside the EU in Australia, India, Indonesia, Italy, New Zealand, Portugal, Uruguay and the USA. WEGO-ITN is positioned in different disciplines (political ecology, feminist studies, human geography, anthropology and critical development studies) and has explored a wide range of topics (as reflected in the book) from extractivism, commoning, care, communities, livelihoods, embodied subjectivities to degrowth resistances to mainstream economic development. The network aimed to do self-reflective and non-extractive feminist research, working with individuals, local communities and social movements, contributing to FPE debates in and out of academic spaces.

In its approach, WEGO-ITN took up a grounded understanding of FPE to look at how everyday practices of social difference, environmental change and political economies across scales in 'a process of doing environmentalism, justice and feminism differently' (Harcourt & Nelson, 2015, p. 9). WEGO-ITN's emphasis has been to undertake research and other activities that discern social and ecological injustices at the intersection of systems of oppression (patriarchy, race, class, coloniality, speciesism) and collectively build equitable social and ecological transformation (Cho et al., 2013; Elmhirst, 2018). As the network has recorded on its website, WEGO-ITN has explored the multiple forms of knowledge that shape and co-construct environmental practices and the politics of the everyday through intersectional forms of social difference such as gender, class, ethnicity, age, ability, sexuality, place and nation. In its different projects, WEGO-ITN has raised questions on how nature, culture and society co-constitute each other, learning from critiques of science by feminist theories, queer theories and environmental humanities, and decolonial critiques of whiteness and privilege.

In constructing the chapters as dialogues, the book is grounded in the ongoing conversations that made up the collaborative WEGO-ITN research process. The designing, writing and editing of the book was built into the WEGO-ITN training with a series of cross-generational conversations (mostly on-line) among supervisors, mentors and partners, emerging from the 15 PhD projects undertaken by early career researchers. It is important to note that running through the research process are reflections on how COVID-19 changed the nature and focus of WEGO-ITN's research in creative and unexpected ways. Going on-line meant opening up new questions about embodied and in-place convergences and between personal and political space. This posed a challenge in the implementation of feminist methodologies engaged with participatory action research techniques, but it also allowed for creativity to transform the way we employ digital spaces to reach voices far from the places the research is situated. The last years of the pandemic raised diverse questions around languages of care in feminist and environmental justice research, and politics. The encounters with the virus and the isolation it engendered reinforced conversations about how to include more-than-human actors to think with non-western epistemologies, natures and voices.

Reflecting on the complex processes of running a network that was designed for face-to-face connections mainly on-line, the chapters also look at how FPE as a research process based on feminist ethics could respond and adapt to disruptions. The book's conversations move in and out the ethics of doing feminist research aware of historical and contemporary positions of power, paying attention to the authors' shifting positionalities, origins and choices of how to do research. The conversations reflect how it is important to make visible the troubles of doing politically engaged research while learning from the COVID-19 pandemic's restrictions on mobility and face-to-face engagement, as well as the possibilities of using the technical openings in digital space. The chapters illustrate the different ways to produce feminist intersectional and intergenerational knowledge paying attention to research as political and ethical practice online—across geo-political and language differences, differential access to Internet and technical barriers, etc. The book therefore features a range of styles reflecting different possibilities: each of the authors had to write and reflect from wherever they had landed during the collaborative process. Such an open approach has helped to create new methodological, theoretical and epistemological ways of doing research across geographical arenas, breaking down older barriers around needing to travel and be in-place. As a result, most of the writing is collaborative and fluid, allowing for reflective, emotional and creative responses to the thorny questions being asked around power, resistance and pain, with some chapters using art, photos, pictures as well as storytelling.

By showcasing the experiential and emotional nature of doing FPE research, the book illustrates the continuous praxis of feminist knowledge production and how it is built through relationships in specific sites, including in digital space. By using the form of conversation and making transparent the positioning of the authors, the book's methodology is itself a critique of normative scientific models that are based on the idea of the rational and disembodied researcher. The choice of writing conversations is also a way to keep enquiries open, as the chapters do not attempt to resolve the questions, but to probe and highlight diverse perspectives, contradictions, self-reflexivity, as well as the complexities of intergenerational and intersectional researcher–participant relations. Storytelling is the vehicle we use to recognise the plurality of ways of producing knowledge, such as orality, and distances us from modern hegemonic stories and rational economic narratives that support the existence of a unique and absolute truth. In this way, the book embraces feminist research as a continuous learning and unlearning process.

The book acknowledges the productive tension around putting together a collaborative feminist book positioned on the margins of academe, yet also funded by the EU and hosted by a whole set of academic practices, including consolidating careers and gaining a PhD. The chapters navigate the uneasiness of the funding/research/praxis, with the political aim to do societally relevant and meaningful non-extractive research. Each chapter contains a back and forth between method and theory, between the known and unknown, what counts as knowledge and data and how to respect 'otherwise' logics and world views. While the chapters on decoloniality address some of these concerns most directly, each chapter reflects on what counts as knowledge and whose knowledge we are able to share, concerns which deeply informed the WEGO-ITN project.

Themes Emerging from the Book's Conversations

As the title 'Contours of Feminist Political Ecology' suggests, the book aims to sketch out a series of common debates in feminism and political ecology, re-opening and troubling themes such as population (Chapter 10), extractivism (Chapters 2 and 3) and more-than-human relationalities (Chapter 6), while also touching on topics that are not often unpacked or mobilised in political ecology, such as ageing (Chapter 5), health and the body (Chapters 4 and 12) and the use of art in research and activism (Chapters 3 and 11). The book also revisits central themes in feminist debates, such as care (Chapters 7 and 8) and decoloniality (Chapters 9 and 12) through a political ecology viewpoint. Finally, it engages with plural imaginings and futures of socio-ecological transformations, informed by critical feminist thinking. Importantly, the chapters in this book engage with these themes and topics in a dialogical and open-ended way, shaping together contours of FPE that are intentionally and explicitly porous and incomplete. The stories told in many of the chapters often originate from different geographies and include distinct identities, histories and place-based complexities. Going beyond a mere comparison between such cases, each chapter presents these stories as opportunities to reflect together on the analytical and theoretical frameworks that link them together and which emerge from our ongoing dialogue and interactions as researchers in the WEGO network.

A common thread running through the chapters is the effort made by the authors to situate ourselves in the writing process, as we write stories about places we visited (and/or inhabit), people we interact(ed) with, communities we learn with and systems of power and oppression that touch upon us in very different yet very connected ways. This is expressed not so much in dry declarations of our nationality, profession or skin colour, but more in terms of autobiographies of pain and illness connected to place, gender and race (see Chapter 4), intimate and embodied understandings of theory such as age (Chapter 5) and decoloniality (Chapters 9 and 12), honest sharing of frustrations and fears over topics such as population growth and population control (Chapter 10), and the placing of personal memories of place and more-than-human connections under an analytical lens of subjectivity formation (Chapters 6 and 12). All chapters strive to remain aware of how the knowledge we produce is necessarily tied to our own histories and positionalities.

A key theme and point of reflection in many of the chapters are the interrogation of knowledge politics and how such politics are often tied to colonial and patriarchal histories and legacies. Dismantling such ways of thinking around 'natural resources' and broader ecologies, including those of human bodies, as well as ways to confront the climate crisis, or ways of living that provide meaning and well-being to communities, is a question engaged with from different starting points and geographies. In Chapter 2, authors reflect on the COP26 process and the dominant discourse perpetuating climate colonialism in these arenas, underpinned by processes of capitalism, imperialism and development. They counter-pose these dominant notions of climate denialism, greenwashing and scripts for unjust transitions, including Net Zero, with the intimate and untold stories of oil palm and extractivism in Indonesia, environmental change and climate impacts in pastoralist communities in Kenya and resistance to onshore oil extraction in the UK. Through their analysis, they show how extractivism and climate coloniality take form and operate in the everyday, including in false solidarities, but also how communities articulate alternatives that subvert such narratives in practice. Chapter 3 continues that interrogation, reflecting on the process, experiences and learnings of co-curating a series of exhibitions and events as part of a feminist ecological politics. The authors look at how such experiences cultivated care, foregrounded community and wove connections between extractive contexts. They pay close attention to how the emotional and the embodied enter scholar-activist exhibition spaces as a way to resist extractivism and find solidarity. Chapter 4 engages with the themes of porosity and embodiment to trouble knowledge produced and reproduced around health and bodies in relation to local ecologies and understandings of place through the narration of experiences around endometriosis, malaria and hyperemesis gravidarum in Southern Europe and the USA. Authors show the political ecological nature not only of the occurrence of illness, but also of the discourse generated around it and its causes and its risks. The authors point to the reproduction of intersectional injustices as well as ways to resist them, involving the use of certain technologies.

Pushing FPE thinking on injustices further, Chapter 9 unpacks and challenges the Eurocentric context of meaning making with reflections on the authors' epistemic relationship with coloniality. They share with the reader their place-based position as researchers from the 'global' South, defining their research as an outcome of thinking through decoloniality. They explore how FPE and decoloniality give diverse meanings to subjectivity, the body and the other as they outline how decoloniality informs FPE research. The dialogue with FPE and decoloniality is continued in Chapter 12, which looks at how meanings flow between Western and Andean cultural horizons in urban environments marked by coloniality. The authors share elements from the Andean worldview from a decolonial perspective, inviting readers to consider Andean ways of understanding and feeling the relationship between the social and the natural as ways to learn about how to cope with social and environmental crises otherwise.

Continuing the engagement with open, pluriversal approaches to ecologies, Chapter 11 looks at ways to shape social and environmental futures not from a top-down technical or policy perspective but through a grounded engagement in the imagination and vision of market vendors in Mexico City. The author addresses energy in the future, learning from market women's dreams and fears in order to arrive at otherwise ways to address energy transitions, unpacking the implications these visions may have for communities and environments. The chapter illustrates a feminist methodology that uses storytelling to build knowledge, working with communities' own visions of their futures.

While the overall tenor of the book intends to be open, learn from difference and think positively about possibilities for the future, it is also evident that the contours of FPE have been shaped by contestation. Paying attention to what is uncomfortable and where emotions are sparked informs FPE as a vibrant and dynamic set of discourses. In Chapter 10 the authors candidly look at the difficult (and often avoided) topic of population in FPE reflecting on the responses to Donna Haraway's (2016) call to 'make kin not babies'. They speak of others (and their own) strong, even explosive, reactions to Haraway. The informal dialogue format allows the authors to map out, from both a personal and scholarly level, the contours of feminist thinking about the fraught topic of population growth and population control, using emotions as an entry point into academic and political debate.

Another strong and emotive theme throughout the book is the importance of care and its different (and contested) meanings in FPE theory and practice. Chapter 7 features an intergenerational discussion on meanings of care looking at care in relation to ethics, intersectional justice, feminism and environmental activism. It highlights how care in activism, teaching and research is part of FPE approaches to knowledge and politics from both a Global South and North perspective. The activist scholars in Chapter 8 look at how care is central to the move towards radical social, economic and environmental change in their reflections on FPE contributions to degrowth as an academic and activist movement. The authors converse on FPE perspectives on care with a feminist critique of the structural racial, gender and wider social inequalities perpetuated by growth-dependent economic systems. They consider, with examples from their own experiences with communities in both the Global South and North, how paying attention to care work and valuing care for humans and more-than-human others is crucial to building just, sustainable and convivial societies. A practical example of how care informs FPE practice and research is illustrated in Chapter 3. This chapter foregrounds how the authors cultivated care as they worked with communities to create an exhibition, paying attention to how the emotional and the embodied enter into scholar-activist encounters. Chapter 5 also shows how care is part of FPE theorising and practice in an examination of ageing experiences as relational processes that require an awareness of how the ethics of care is embodied in everyday practices. In the authors' practice and research on ageing in two very different contexts, they analyse how ageing is part of dynamic socionatural relations, arguing that FPE needs a more careful awareness of ageing experiences.

Another important theme that flows through the book is relations with the more-than-human.

Chapter 6 explores the politics of interdependencies through situated entanglements with water. The authors set out how more-than-human interdependencies within FPE means starting from an understanding of relationality. Drawing on research with waters and communities in Maharashtra, India and the Tagus River in Spain, the chapter examines the co-constitution of embodied subjectivities with the more-than-human, addressing issues of well-being, illness and ecological change in contemporary waterscapes. The chapter's discussions on the contradictions, tensions and ethical implications of situated more-than-human co-becomings is echoed in the dialogical debate around Andean ways of understanding territory in Chapter 12 and the tensions with a modern understanding of urban landscapes, which see nature as something to be exploited and extracted, erasing other culture's meanings and understandings of living with the more-than-human.

These are some of, but by no means all, the main themes that inform how FPE emerges in the book. By highlighting multiple perspectives, contradictions, self-reflexivity, as well as the complexities of intergenerational and intersectional researcher-participant relations, the book shows how feminist research is a continuous learning and unlearning process. This includes navigating tensions between funding, research and praxis, method and theory, the known and unknown and sensing and feeling (*sentipensar*). Throughout the book the authors grapple with the tensions of doing research that meets the requirements of research institutions (the university space, the EU), while respecting different logics and world outlooks. There is a sense of excitement in doing FPE research which, as the different chapters show in creative and unexpected ways, opens questions and makes convergences in fields not previously considered. We hope that the multi-layered debate in the chapters around the diverse meanings, understandings and languages of FPE invites readers to think further about how to continue to shape the contours of FPE.

Where to Go from Here?

The book has set out how WEGO-ITN[1] has evolved through meaningful and careful research as the world faces ongoing climate crisis, the prospect of future pandemics, wars, economic and political uncertainty and reversals on gains made, including in the areas of gender equality, intersectional justice and human rights. We hope the book contributes to an understanding of how FPE can support our individual and collective resilience in the future.

The book illustrates the importance of innovation and adaptation to becoming experimental, creative and flexible in order to deal with individual, institutional and global uncertainties. It sets out how we are learning to cope in this new normal: pandemics, economic and social uncertainty and climate crisis. The quest is how to build relationships both virtually and in-place; politically and culturally that can shape FPE as a space for creative learning. We hope to 'live the talk' of a feminist network of transformation that centres relations of care as we look for alternatives to capitalism and mainstream development through engagement with communities.

This relational approach is a key dimension of FPE. While the hegemonic patriarchal mode of development is based on domination, conflict and exploitation (over bodies, cultures, nature), FPE promotes a transition based on the day-to-day practices of women, men, others and their communities to sustain ecologically viable livelihoods. The shaping of these livelihoods takes place within the tension between autonomous and diverse imaginaries and the impositions of capitalist globalisation. Imagining the lives and the worlds individuals and communities want to live in needs to happen outside the development discourse that has already imagined and determined the world for all of us. We hope the reader

[1] WEGO-ITN has transitioned into the FEST* network (Feminist Ecological Solidarities for Transformation) network. The direction for the FEST* network will be to continue FPE conversations expanding to different spaces undertaking activist research with people engaged in intersectional intergenerational environment justice in communities/institutional arenas and bring their stories and strategies together in a series of FPE dialogues, workshops and other creative encounters. WEGO-ITN held five FPE national dialogues between July 2021 and May 2022 in Spain in dialogue with Latin America, The Netherlands, Germany, Italy and the UK. These FPE dialogues were set up in order to contribute not only to academic debates but also to societal and institutional change highlighting the importance of community campaigns around gender and labour rights, food sovereignty, well-being, body politics, ageing and rights to health and clean water.

agrees that the book opens future possibilities on the basis of our collective research and encounters; our experimental and culturally anchored practices aimed at putting Life at the centre.

Funding: This chapter was funded by the Wellbeing Ecology Gender and cOmmunities Innovation Training Network (WEGO-ITN) funded by the European Union's Horizon 2020 research and innovation programme under the Marie Sklodowska-Curie grant agreement No. 764908-WEGO 2018-2021.

References

Agarwal, B. (1998). Environmental management, equity and ecofeminism: Debating India's experience. *The Journal of Peasant Studies, 25*(4), 55–95.

Agostino, A. (2018). Towards an urban agenda from a Feminist Political Ecology and care perspective. In C. Bauhardt & W. Harcourt (Eds.), *Feminist political ecology and the economics of care: In search of economic alternatives* (pp. 262–286). Routledge.

Barca, S. (2020). *Forces of reproduction: Notes for a counter-hegemonic anthropocene*. Cambridge University Press.

Candraningrum, D. (2018). The spiritual politics of the Kendeng Mountains versus the global cement industry. In K. Worthy, E. Allison, & W. A. Bauman (Eds.), *After the death of nature: Carolyn Merchant and the future of human-nature relations* (pp. 219–234). Routledge.

Cho, S., Crenshaw, K., & McCall, L. (2013). Toward a field of intersectionality studies: Theory, applications, and praxis. *Signs: Journal of Women in Culture and Society, 38*(4), 785–810.

Dengler, C., & Seebacher, L. M. (2019). What about the global south? Towards a feminist decolonial degrowth approach. *Ecological Economics, 157*, 246–252.

Di Chiro, G. (2017). Welcome to the white (M)anthropocene?: A feminist-environmentalist critique. In S. Macgregor (Ed.), *Routledge handbook of gender and environment* (pp. 487–505). Routledge.

Elmhirst, R. (2011). Introducing new feminist political ecologies. *Geoforum, 42*(2), 129–132.

Elmhirst, R. (2018, January 10). Ecología política feminista: Perspectivas situadas, compromisos emergentes. *Ecologia Politica*, 54. Retrieved August 27, 2022 from https://www.ecologiapolitica.info/ecologias-politicas-feministas-perspectivas-situadas-y-abordajes-emergentes/

Fanon, F. (2001). *The Wretched of the Earth*. Penguin Modern Classics.

Federici, S. (2004). *Caliban and the Witch: Women, the body and primitive accumulation*. Automedia.

Fraser, N. (2016). Contradictions of capital and care. *New Left Review, 100*, 99–117.
Gibson-Graham, J. K., Cameron, J., Dombroski, K., Healy, S., & Miller, E. (2020). Cultivating community economies: Tools for building a livable world. In J. G. Speth & K. Courrier (Eds.), *The new systems reader* (pp. 410–432). Routledge.
González-Hidalgo, M., & Zografos, C. (2020). Emotions, power, and environmental conflict: Expanding the 'emotional turn' in political ecology. *Progress in Human Geography, 44*(2), 235–255.
Gururani, S. (2002). Forests of pleasure and pain: Gendered practices of labor and livelihood in the forests of the Kumaon Himalayas, India. *Gender, Place and Culture: A Journal of Feminist Geography, 9*(3), 229–243.
Haraway, D. (2001). *Simians, cyborgs, and women: The reinvention of nature.* Routledge.
Haraway, D. (2016). *Staying with the trouble: Making kin in the Chthulucene.* Duke University Press.
Harcourt, W., & Nelson, I. L. (Eds.). (2015). *Practicing feminist political ecology: Moving beyond the green economy.* Zed Books.
Harding, S. (1991). *Whose science? Whose knowledge?: Thinking from women's lives.* Cornell University Press.
Harris, L. (2015). Hegemonic waters and rethinking natures otherwise. In W. Harcourt & I. L. Nelson (Eds.), *Practicing feminist political ecologies: Moving beyond the green economy* (pp. 157–181). Zed Books.
Kothari, A., Demaria, F., & Acosta, A. (2014). Buen Vivir, degrowth and ecological Swaraj: Alternatives to sustainable development and the green economy. *Development, 57*(3), 362–375.
Liboiron, M. (2021). *Pollution is colonialism.* Duke University Press.
Lugones, M. (2010). Toward a decolonial feminism. *Hypatia, 25*(4), 742–759.
Mellor, M. (2006). Ecofeminist political economy. *International Journal of Green Economics, 1*(1–2), 139–150.
Mohanty, C. T. (2003). *Feminism without borders: Decolonising theory.* Duke University Press.
Mollett, S. (2017). Gender's critical edge: Feminist political ecology, postcolonial intersectionality, and the coupling of race and gender. In S. MacGregor (Ed.), *Routledge handbook of gender and environment* (pp. 146–158). Routledge.
Mollett, S. (2018). The Río Plátano Biosphere Reserve: A postcolonial feminist political ecological reading of violence and territorial struggles in Honduras. In S. Mollett & T. Keep (Eds.), *Land rights, biodiversity conservation and justice* (pp. 184–205). Routledge.
Mollett, S., Vaz-Jones, L., & Delicado-Moratalla, L. (2020). Feminist political ecologies: Race, bodies and the human. In A. Datta, P. Hopkins, L. Johnston,

E. Olson, & J. M. Silva (Eds.), *Routledge handbook of gender and feminist geographies* (pp. 271–284). Routledge.

Nightingale, A. J. (2003). A feminist in the forest: Situated knowledges and mixing methods in natural resource management. *ACME: An International Journal for Critical Geographies, 2*(1), 77–90.

Nightingale, A. J. (2011). Bounding difference: Intersectionality and the material production of gender, caste, class and environment in Nepal. *Geoforum, 42*(2), 153–162.

Nirmal, P., & Rocheleau, D. (2019). Decolonizing degrowth in the post-development convergence: Questions, experiences, and proposals from two Indigenous territories. *Environment and Planning E: Nature and Space, 2*(3), 465–492.

Ojeda, D., Nirmal, P., Rocheleau, D., & Emel, J. (2022, August 1). Feminist Ecologies. *Annual Review of Environment and Resources* 47 (Volume publication date October 2022) Review in Advance. https://doi.org/10.1146/annurev-environ-112320-092246

Puig de la Bellacasa, M. P. (2017). *Matters of care: Speculative ethics in more than human worlds*. University of Minnesota Press.

Quijano, A. (2000). Coloniality of power and Eurocentrism in Latin America. *International Sociology, 15*(2), 215–232.

Resurrección, B. P. (2017). Gender and environment from 'women, environment and development' to feminist political ecology. In S. MacGregor (Ed.), *Routledge handbook of gender and environment* (pp. 71–85). Routledge.

Rocheleau, D. (2015). A situated view of feminist political ecology from my networks, roots and territories. In W. Harcourt & I. Nelson (Eds.), *Practising feminist political ecologies: Moving beyond the Green economy* (pp. 29–66). Zed Books.

Rocheleau, D., Thomas-Slayter, B., & Wangari, E. (1996). *Feminist political ecology: Global issues and local experience*. Routledge.

Seager, J. (1993). *Earth follies: Feminism, politics and the environment*. Routledge.

Shiva, V. (1988). *Staying alive: Women, ecology, and survival in India*. Kali for Women.

Singh, N. M. (2013). The affective labor of growing forests and the becoming of environmental subjects: Rethinking environmentality in Odisha, India. *Geoforum, 47*, 189–198.

Sultana, F. (2011). Suffering for water, suffering from water: Emotional geographies of resource access, control and conflict. *Geoforum, 42*(2), 163–172.

Sultana, F. (2021). Political ecology 1: From margins to center. *Progress in Human Geography, 45*(1), 156–165.

Sundberg, J. (2004). Identities in the making: Conservation, gender and race in the Maya Biosphere Reserve, Guatemala. *Gender, Place & Culture, 11*(1), 43–66.
Sundberg, J. (2017). Feminist political ecology. In D. Richardson (Ed.), *The international encyclopedia of geography*. Wiley-Blackwell & Association of American Geographers.
Tronto, J. C. (2015). *Who cares?: How to reshape a democratic politics*. Cornell University Press.
Velicu, I. (2015). Demonizing the sensible and the 'Revolution of Our Generation' in Rosia Montana. *Globalizations, 12*(6), 846–858.
Wichterich, C. (2015). Contesting green growth, connecting care, commons and enough. In W. Harcourt & I. L. Nelson (Eds.), *Practising feminist political ecologies: Moving beyond the "green economy"* (pp. 67–100). Zed Books.

CHAPTER 2

Untold Climate Stories: Feminist Political Ecology Perspectives on Extractivism, Climate Colonialism and Community Alternatives

Dian Ekowati, Siti Maimunah, Alice Owen, Eunice Wangari Muneri, and Rebecca Elmhirst

Prelude

Notes from COP26 Climate Conference: Confronting Climate Coloniality

We begin this chapter with our reflections from the United Nations COP26 climate conference (26th meeting of the Conference of Parties on climate) in late November 2021, following our participation in various

D. Ekowati · A. Owen · R. Elmhirst (✉)
University of Brighton, Brighton, UK
e-mail: r.j.elmhirst@brighton.ac.uk

A. Owen
e-mail: A.Owen@brighton.ac.uk

W. Harcourt et al. (eds.), *Contours of Feminist Political Ecology*,
Gender, Development and Social Change,
https://doi.org/10.1007/978-3-031-20928-4_2

events in and outside the summit. The summit itself was a space dominated by corporations, fossil fuel companies and powerful governments and included a zone for NGOs, civil society groups and green capitalism entrepreneurs promoting carbon reductionism (Net Zero), the financialisation of nature and a raft of techno-utopian innovations. Politicians and corporate leaders brought activist slogans into their speeches, while youth and civil society voices were celebrated and 'staged' within the mainstream conference (Aykut et al., 2022). Outside COP26 itself, a 'fringe' of loosely connected activist spaces were sites of counter-narrative and the expression of decolonial, anti-colonial, anti-racist and feminist politics. The Peoples' Summit involved social and environmental justice activists, youth, Indigenous groups, critical academics and trade unions whose diverse registers and claims converged around a sense of grassroots globality in opposition to the re-enchanting of green capitalism that was going on inside the COP (Aykut et al., 2022).

From our vantage point as feminist political ecology (FPE) researcher-activists experiencing the dissonances, exclusions and erasures as we navigated these COP26 spaces, we witnessed the contrast between the climate narratives of the corporate fixers and the stories from those who embody the impact of extractivism and corporate greed: mining, toxic waste, oil drilling and ecological degradation. Our COP26 reflections inspire the questions that frame our dialogue in this chapter: how does climate colonialism surface in the reflections we share from our research and activism? How is it that the root causes of climate change—extractivism, injustice and disconnection from nature—are simultaneously rendered invisible and reinforced in corporate and state responses? Are colonial and extractive injustices being reproduced in green initiatives as nuanced community perspectives remain unheard? What alternatives to extractivism might be heard in these stories?

Mai: I am reminded of Indonesian President Jokowi's speech delivered at the World Leader's Summit at the beginning of COP26 in Glasgow, UK. He put "climate change as a major threat to global prosperity and

S. Maimunah
University of Passau, Passau, Germany

E. Wangari Muneri
Institute of Development Studies, University of Sussex, Brighton, UK

development", and then went on to answer that threat by offering, among other things, the development of an 'electric car ecosystem', and the development of a green economy-style Special Economic Zone (SEZ). All this means that Indonesia admits to the existence of climate change but continues extractive development with the same pattern. The expansion of the electric car industry means the enlargement of forest and land clearing for extracting nickel in islands of Sulawesi and Maluku, and coal extraction in Kalimantan to produce the materials and dirty energy for the manufacture of electric car batteries. The expansion of the biofuel economy means the expansion of large-scale palm oil plantations in remote parts of the country, especially in Papua. The proposed measures also include a green SEZ that involves constructing a large-scale dam along the Kayan river to power a smelter plant in Kalimantan. Isn't his speech a form of climate denial?

Eunice: My experience of COP 26 was different from any other I had previously attended. Not only because I attended virtually, but the attendance by lobbying "polluters" like giant fossil fuel companies and social justice activists had increased significantly compared to previous years, warranting mention by various media outlets. The overwhelming presence of the trending #NetZero on LinkedIn and the notable attendance of lobbyists from the fossil fuel industry confirmed COP as a pledging event, which codifies greenwashing. I got the sense that the Peoples' Summit had become larger and more organized, which is a ray of hope that the greenwashing happening on the inside is being countered. Like previous COPs, there was lack of nuance; an example is the classification of livestock being bad for the environment and climate. The issue was picked up by other scholars who brought sheep to the COP with the aim of illustrating that the issue is not livestock per se, but their intensified production. This message resonated with the Maasai community that I work with in Kenya, and how they are disproportionately affected by extreme weather patterns but now risk facing social penalties by being lumped in with intensive livestock farmers elsewhere in the world who are the targets of climate activism.

Dian: I presented at the COP Coalition Peoples' Summit alongside the other authors here. I presented the stories of different people in Kalimantan, Indonesia, whose life has been changed by oil palm in different ways, depending on their class, ethnicity, migration status, ability, age and gender. There are terrible stories and there are some hopes. There are differences in ways oil palm is produced, and differences in ways it affects peoples' life around it. I remembered that one of the participants in our COP Coalition event from the Global North expressed their shock as they think of oil palm production and its impacts in a homogeneously violent way, and another participant expressed concern about greenwashing in oil palm large-scale sustainability programmes, both coming from the belief that boycotting palm oil will do good for all. At the same time, I remember the anxieties of some small-scale oil palm farmers I met as boycotting destroys their hope and reduces the price for their oil palm harvest. The green discourse in the Global North towards oil palm, that often overlaps with the movement to support Indigenous people, brings different impacts to those rural people who do not necessarily fall into the 'Indigenous' category and who interact with oil palm trees in their everyday life. And this impact is not always a good thing.

Alice: Throughout the COP, the side events and the People's Climate Summit different people and organisations created and held diverse spaces for storytelling, sharing and listening. Through these stories and spaces, diverging ways of knowing, ways of doing things and ideas about climate action were articulated and imagined. Striking juxtapositions arose. Many concerned with climate justice are urging deep unlearning from our histories, learning the histories of colonisation and discovering languages, practices, stories which have been marginalised and which might be revived. At an activist event, an Aboriginal speaker (from colonised Australia) described how, in their culture, they are "walking backwards" into the future, looking back towards their ancestors with 60,000 years of land stewardship and harmony rather than speculating about the future. Conversely, at an unofficial side event the next day with 'storytelling' in the description, a group of mostly Australian financiers and bankers shared their 'visions' for the future; their pragmatic imaginations conjured scenes of what Net Zero 2050 would look like in ways that I'm sure they thought were utopian (international investment innovations, green industry) but to me seemed to be taking us closer towards dystopia. "If we believe it, it will come true" they said with conviction, and it scared me to think they might be right as their speculations begin to materialise from these sectors so detached from what it means to be good stewards of the land.

Inspired by our reflections on COP26 we weave a dialogue through stories and reflections from our research and/or activism in Indonesia, Kenya and the United Kingdom. We explore what is learned when our reflections on grand narratives and systemic injustices are woven together in the warp and weft of feminist political ecology, with its emphasis on situated knowledges, lived experience and the everyday. Our various threads converge and diverge around the issues that decentre root causes, erase nuance and extend injustices in climate responses.

Introduction

The story of climate breakdown responses is one of dissonance between mainstream discourses that highlight capitalist market- and techno fixes and those of climate justice activists whose counter narratives call these out as 'false solutions' that perpetuate injustices and fail to address root causes. In this chapter, we respond to what Farhana Sultana has described as the 'unbearable heaviness of climate colonialism': an ongoing coloniality underpinned by processes of capitalism, imperialism and development that were inherent in the staging of COP26 (Sultana, 2022, p. 3). We do this by bringing together our reflections—what we have previously labelled as 'untold stories'—from our research and activism with communities in Indonesia, Kenya and the United Kingdom.[1]

Dian Ekowati reflects from her research on the everyday care that enables life to be sustained in the oil palm landscape, a landscape that is often described as an extractivist agricultural system, but that at the same time is framed as a green alternative to the carbon economy. Siti Maimunah is an activist and researcher working alongside communities in Kalimantan, but more broadly is seeking to understand the operation of resource extractivism in Indonesia, working with NGOs to support communities affected by extractivism projects. Alice Owen brings her insights from her research on the politics of knowledge as a campaign against a new onshore oil extraction site (Horse Hill) has unfolded in the South of England. Although the local impacts of the site can seem unspectacular compared to extractivism elsewhere (including by British companies in areas colonised by Britain), critical attention is drawn to the climate impacts of the project through campaigning and protest. This exposes the ways in which extractivist logics and implicit climate denial permeate the local experience and the global climate crisis. Eunice

[1] We refer here to our presentation at the People's Summit for Climate Justice in Glasgow, UK from 6 to 10th November 2021. Talks from the event have been lodged here: https://cop26coalition.org/talk/.

Wangari reflects from her research on the gendered nature of environmental changes in the Maasai pastoralist community in Kenya as the community adapts to erratic weather events like prolonged droughts and recurrent floods, and where relations between people and cattle create differentiated experiences to environmental changes. Rebecca Elmhirst contributes through her immersion in these stories and the questions they inspire and address, alongside the reflections she brings from longstanding research with communities in Indonesia's oil palm landscapes. Three of us write from positionalities embedded in the Global South, and two of us in the Global North.

We build our dialogue around the systemic roots of the climate crisis and unjust responses, understood through the concept of extractivism (Willow, 2018). Extractivism refers to an increasingly prominent modality of capitalist accumulation based on destructive processes of subjugation, depletion and exploitation of nature and life. It includes the exploitative extraction of a broad range of natural and human resources from colonies and ex-colonies in Africa, Asia and the Americas (Veltmeyer & Petras, 2014) and as such, is deeply entwined with the dynamics of coloniality, imperialistic forms of corporate power and deepening inequalities (Pereira & Tsikata, 2021). Its logic is one of endless growth, corporate enclosure of land and water, erosion of biodiversity and the exploitation of life, rooted in and enabled by coloniality (Gómez-Barris, 2017).

Our dialogues draw on a loosely convened feminist political ecology (FPE)—a nexus of environmental feminisms based around an understanding of and response to global systems and their material consequences (Sundberg, 2016). Bringing feminist political ecology perspectives to bear on extractivism means we connect an analysis of global systems with lived experience, the everyday, the emotional and personal, and do these by attending to intersecting forms of power, including patriarchy, racism and coloniality, worked through at multiple scales. Our research reflections do not share a common conceptual framework, but our connections in the convening space of FPE means we share an understanding of extractivism as configured differently in places with particular histories, relationships and responses to the logics of racial capitalism (Bhattacharyya, 2018; Gómez-Barris, 2017; Pereira & Tsikata, 2021), and this is reflected in the stories we explore in the chapter.

We have previously described these stories as 'untold' in the sense that the coloniality of climate knowledges and discourses marginalises and eclipses those stories that do not easily fit into mainstream climate policy narratives (Chao & Enari, 2021). Moreover, where stories are incommensurate and incompatible with the kinds of stories that ignite globality and connection in climate activism, they may remain unheard, risking new and perhaps hidden forms of injustice for the most marginalised in marginalised communities. FPE requires an ethics of care in how we theorise, research, discuss and write, attuned to the diverse, situated and nuanced ways in which each of us knows, recognises and embodies intersecting forms of power. Our reflections are partial, dependent on our positionalities and our geographical, ontological and political situatedness.

Rather than bringing ventriloquised narratives 'from' communities, we build dialogues around our reflections, shaped through the myriad relationships of academia, activism, friendship and family, in which each of us is enmeshed. Donna Haraway (2019) states that it matters what stories tell stories, it matters whose stories tell stories. We consider what we learn when we bring our reflections from our research and activism contexts together and suggest that in relating/re-telling together, the adjacency of differently situated stories posed through some common themes helps gain perspective on the contours of extractivism and climate coloniality in the everyday and helps us to confront the challenges of bringing nuance and avoiding erroneous solidarity when these stories are brought into climate justice activism spaces. We begin by outlining what we mean when we describe extractivism as a root cause of climate coloniality in the contexts of our research, before opening our dialogue around questions that emerge when we bring our stories together. We close with some reflections on community alternatives, where these foster or recover more reciprocal ways of living outside an extractivist logic.

Climate Coloniality: Extractivism as Root Cause

In early 2022, several months after COP26, the Intergovernmental Panel on Climate Change's working group 2 published its report on the impacts of climate change on people. Significantly, and for the first time, the report named colonialism as a driver of the climate crisis and as an ongoing issue that is exacerbating community climate vulnerability (IPPC, 2022). This attention to root causes associated with an ongoing coloniality in climate impacts and responses has long been a connecting thread linking campaigns for climate justice, including those of Indigenous and land-based peoples' movements in the Global South, race and environmental justice activists and those rooted in global justice/anti-poverty campaigns. Climate justice aligns with the environmentalism of the poor and dispossessed in their struggles for land and livelihood and against an underlying economic system that provides profit for a few but depletion and harm for many (Tokar, 2020).

When we describe extractivism as a root cause of climate harm, we are referring to what Chagnon et al. (2022, p. 3) describe as 'a complex of self-reinforcing practices, mentalities, and power differentials underwriting and rationalising socio-ecologically destructive modes of organising life through subjugation, depletion, and non-reciprocity'. While there has been some debate that the concept should be restricted to its origins in the white settler context of Latin American political economy, core–periphery inequalities and mineral extraction at scale, we follow a more expansive reading of extractivism located in feminist political ecology. We see the logic of extractivism as involving the appropriation of human and more-than-human life forms, depleting and draining in a potentially irreversible way. Extractivist logics involve a centralisation of power and the deepening of relational and intersecting inequalities (colonial, racial, patriarchal, interspecies): extractivism is a modality of 'development' that conditions and pressures all life forms (Chagnon et al., 2022). A range of experiences across different geographical settings exists within these abstract characterisations of extractivism, reflecting different histories of coloniality and settlement, contemporary geopolitics and the 'nature' of what is being extracted.

On the Indonesian island of Kalimantan, the ecological injustices produced through extractivism are clear, taking shape through the country's mercantilist colonial history and its legacies, and through more regionalised forms of political dominance and oppression between dominant and marginalised ethnicities (Chua & Idrus, 2022). Since the colonial period, East Kalimantan has been an area for land, forest, mineral and oil exploitation, with rivers turned into the transportation infrastructure for the extraction of raw materials to supply markets in the Global North. Coloniality is expressed in the state's granting of land concessions to international mining companies: like the colonial strategy of *divide et impera*, these mining companies divide farmers and ecosystems to get 'cheap nature' (Moore, 2015). Siti Maimunah (Mai) describes the depletion and draining of life as extractivist mining, and its irreversible impacts make other ways of relating to land impossible:

Mai: I received a series of news reports via WhatsApp from a friend from East Kalimantan when I had just arrived in Scotland to attend COP

26 in Glasgow. Febi Abdi, 25 years old, was a resident of Makroman village, Samarinda, East Kalimantan, Indonesia. His body was found at an abandoned coal pit owned by PT Arjuna. Febi is at least the 40th person to die in an abandoned mine pit in the last decade; most victims were children. Nearly 60% of Indonesia's 3,033 abandoned mining pits are in East Kalimantan and most are coal pits (Shahbanu et al., 2018). In Makroman village, mining companies managed to coerce the local farmers in the hilly areas into selling their land for mining extraction, with the promise of employment with the company. In the growing seasons that followed, rice fields (sawah) on the lower slopes began to fail due to erosion, flooding, and water shortages. Surviving sawah owners must struggle with the rise in production costs due to reduced soil fertility, water shortages, and weed and pest attacks, eventually selling their formerly productive fields to the mining companies with a loss of livelihood.

In Murung Raya, Central Kalimantan, the operation of an extractive mega project affected the identity of the Murung people and changed their relationship with nature. The river was the primary source of clean water, while river fish and game meat from the forest was the primary protein source. Today, the river water is no longer drinkable, and the Murung have to collect water from creeks far from the village or buy it. Coal extraction in Kalimantan is a history of capital accumulation and destruction of the earth that has contributed to the pollution of the atmosphere. As this damage is ignored, extractivism contributes to the accumulation of climate disasters globally.

Elsewhere in Kalimantan, the logic of extractivism takes shape through the granting of oil palm concessions to large-scale companies. Corporate investments in oil palm in Indonesia exhibit the key features of agrarian extractivism: vast, capital-intensive monocultures reliant on external inputs and technologies, driven by profit-maximisation rather than social and ecological well-being, creating a form of corporate occupation (Li & Semedi, 2021). The term 'plantationocene' has been used to describe the agro-extractivist mode of production, processing and labour, with attention drawn to the colonial legacies of this way of controlling bodies and nature to enable the extraction of profits (Wolford, 2021). In common with the extractivist logics of mining in Kalimantan, Dian Ekowati describes the impacts of this mode of oil palm production on the communities whose stories she shares in her research:

Dian: The power behind large-scale oil palm grows the trees in such a violent way that depletes lives (human and more than human). Oil palm cultivated on a large-scale covers hundreds of thousands to millions of hectares of land, planted solely in oil palm trees. Monoculture is the only known way for planting oil palm by the oil palm companies. I remember how we often got lost when passing through a company's oil palm blocks (concessions) due to the similarities of surroundings: same trees, pattern of planting and similar dirt roads. Relegating/subjugating other forms of life is the feature of a large-scale oil palm plantation.

The complex entanglements of coloniality and extractivist logics in Kenya reflect colonial legacies that have transformed relationships between people, land and animals. The historical marginalisation of pastoralist communities in Kenya by land dispossession can be traced to the colonial era but continued through the post-colonial administration excising large portions of rangelands and demarcating them for wildlife conservation, separate from humans. In addition to demarcating wildlife conservation parks, further Maasai land was allocated to white settlers and crop growing communities, whom the government perceived as engaging in more profit-oriented forms of production. As Eunice explains, pastoralism was and is still seen as an archaic, primitive and unproductive form of land-use in dire need of modernisation.

Eunice: It is no wonder that pastoralism's negative portrayals dominate pastoral policies. One popular perception was that overstocking and overgrazing of cattle was causing desertification in the fragile drylands. This mistaken view has been refuted in research, but government pastoral policies advocated for sedentarization and restriction of mobility for pastoralists and their livestock with the aim of modernising them to mirror the European livestock farmers in temperate climates.

These more extractive forms of livestock production based on sedentarisation have compromised mobility, a key strategy used by the pastoralists to cope and adapt to climate variability and other changes. Mobility allows for use of spatially heterogeneous and climatically variable resources. Reduction in mobility increases the risks of degradation as only part of the rangeland is heavily utilised, making pastoralists susceptible to droughts.

Eunice: During my field work, elderly respondents recall the earlier days when Maasai grazing land went as far as Nairobi (approximately 250 kms away) and all the way to Laikipia, in Northern Kenya, where the current Northern Maasai are located. They reminisce how access to the rangelands made life easier for the community during events like droughts, floods, pest invasions and diseases. The effects of these land grabs have severely compromised the community's ability to respond to climate shocks and stressors that frequent the region.

Socio-political and colonial processes of agrarian extractivism and neoliberal conservation are transforming landscapes and the ways pastoral communities relate to them, reducing mobility and preventing them from using their traditional knowledge and practices, undermining their livelihoods and life-making.

Our reflections on the specific histories and experiences of extractivism draw out the ways these have been shaped by colonial pasts in the Global South. In what sense is coloniality associated with the extractivist project that Alice is researching in a Global North context and how does this relate to extractivisms elsewhere?

Alice: In school history lessons I learnt about Tudor royalty, the Industrial Revolution and the success of the British Empire, the abolition of slavery. Large, dark chapters of England's history of plunder, violence, dispossession and (cultural) genocide were either neglected or reframed to tell a particular story of the nation's pivotal contribution to global 'progress' and the making of the modern world. Decolonial scholarship and critical histories retell the story of modernity, giving due importance to the colonial encounters by the British and other colonising countries which led to violent erasures and subordinations of peoples, cultures and territories.

An essential tool to colonial expansion and the extraction of wealth were the logics of extractivism, justified through scientific reasoning and Christian morality which compelled the conquering and taming of unruly 'Others'. A key trick underlying the 'Death of Nature' (Merchant, 1980) and in justifying colonial extractivism was (and is) to render the Other as ontologically available for extraction, describing certain peoples as inhuman and other-than-human nature as inanimate. This move transforms the Other into a potential resource, and under the imperative of nation building comes the 'need' to dominate, exploit and accumulate the wealth of these 'resources'. Scientific innovations made (and make) accelerating resource extractivism seem inevitable; the technical ability to map resources pre-empts their extraction, the innovation of technologies aspires to bring them into being. Since 1835 the British Geological Survey has mapped resources in the UK's interests both nationally and overseas. In 2014, the Weald Basin in Southeast England was surveyed by the BGS to estimate the potential shale oil and shale gas resources, indicating between 2.20 and 8.57 billion barrels of shale oil could be in the region. Somewhat inevitably such findings attracted prospective industries, including to the Horse Hill site where both unconventional (shale, requiring additional stimulation such as by hydraulic or acid fracturing) and conventional oil plays have been explored.

The UK and the Industrial Revolution were at the heart of the rise and spread of fossil fuels, setting in motion the climate crisis. The centrality of fossil fuels to industry and society was not inevitable but a choice to maximise the reliability, mobility, productivity and thus profits of industry compared to the use of traditional energy sources such as water mills (Malm, 2016). The rise of urban industry in the eighteenth century was accompanied by the enclosures of the commons, meaning people who had once lived closely with the land with certain rights and responsibilities were forced to find labour in cities. Although much less violent than in colonial contexts, this dispossession and disconnection from the land underpinned by the logics of extractivism also marks a loss of ways of life more in tune with nature. With much industry outsourced from the UK over the last century to countries with less stringent human rights and environmental regulations, it is possible for many in the UK to live without considering either the social and environmental costs of high-consumption lifestyles or the forgotten ways of thinking about nature as something humans are a part of rather than apart from. Perhaps it is through unusual confrontations with extractivism, such as the arrival of potential onshore oil and gas in the English countryside—or indeed experiences of the droughts and heatwaves exacerbated by the climate crisis—that the underlying assumptions of modernity can be brought into question.

In the stories we share, extractivism reflects a political ontology based on imaginaries of human exceptionality, nature–culture dualisms and mechanistic or technocentric ways of understanding or relating to the world. Extractivism extends beyond (ab)using the earth as it is also a way of acting and being in the world; it constitutes a specific way of thinking, knowing and acting—of relating to nature (Willow, 2018), which is normalised in Global North contexts, as Alice describes. Reductive political ontologies underpin green economy initiatives based on achieving Net Zero carbon emissions as a 'solution' to climate change. When green economy initiatives emerge within an extractivist logic, this perpetuates a human mastery of 'nature' through greenwashing technofixes (e.g., offsetting carbon emissions through neoliberal conservation, as described by Eunice) and novel extractions (e.g., palm oil production as biofuel, as described by Dian or rare earth mineral mining, as promoted by Indonesia's president, as Mai has explained). These forms build from and entrench climate coloniality (Sultana, 2022), reinforcing the interplay of colonialism, extractivism and climate injustice. So far, we have considered the ways in which the depleting and draining properties of extractivism unfold in our research contexts. As we thread our way back to the questions inspired in our COP26 reflections, we turn now to explore the themes that emerge when we bring our specific and situated reflections together.

Climate Vocabularies: Expanding the Extractive Frontier

Climate change narratives and accompanying vocabularies are variously mobilised by campaign groups to support claims against extractive projects, sometimes at odds with the concerns and experiences of those experiencing the everyday coloniality of extractivism. Conversely and simultaneously, the state and extractive industries use vocabularies of climate change to legitimise an expanding and deepening of extractivism under the guise of green industry. Our dialogue in this section considers the geographical and discursive dissonance of climate change vocabularies as they are introduced and mobilised in fossil fuel extraction contexts in Indonesia and the United Kingdom.

Climate 'impacts' describe the risks and already unfolding realities of social and ecological breakdowns that result from anthropogenic climate change caused by greenhouse gas emissions. The climate crisis is a planetary phenomenon, but as climate justice campaigns insist, responsibility for and vulnerability to climate change play out along the contours of coloniality and inequality. Climate impact narratives—including in localised and critical analyses of climate impacts—tend not to define the social and ecological impacts of extractivism (the root cause of climate change) as 'climate impacts'. This creates a disconnect between the devastation and violence caused by the *extraction* of fossil fuels (and other socio-ecologically destructive processes) and that which is caused by the *combustion* of fossil fuels. In this sense, we suggest that climate change narratives which describe climate change as happening everywhere or elsewhere can overlook—sometimes strategically, sometimes ignorantly—local everyday experiences of extractivism.

How Is the Vocabulary of Climate Change Dislocated from the Everyday Experiences of Extractivism?

We consider this question from the contrasting fossil fuel extraction contexts of Kalimantan in Indonesia and the Surrey Hills in the United Kingdom.

Mai: In Sungai Murung village, Central Kalimantan, or Makroman village in East Kalimantan, the coal extraction areas on the island of Kalimantan, farmers and women do not use the vocabulary of "climate change" in everyday activities. When I met Tukiyem, a woman vegetable picker, she told me about Genjer leaves (Limnocharis flava—Yellow velvet) which is increasingly difficult to obtain because of coal mining. In Central Kalimantan, while bathing on the Lanting (a floating hut in the river where people bathe and wash clothes while telling stories), I didn't hear anyone talk about climate change. Yet the Lanting reveals the changing nature of flooding. Swidden agriculture (rotational farming) depends on rainwater, but now the rainy season is uncertain. Instead, the women told me a story about women's protests and coal road blockades in 2015 because the river water was polluted by coal mine waste, causing river water to become undrinkable. Coal mines impact rice fields, gardens, and water sources in Makroman village. Meanwhile, Sungai Murung village has been surrounded by logging companies since the 1970s and coal extraction since 2000. Clearing of land, destruction of forests and gardens, use of transportation, and burning of coal are the causes of climate change, destroying nature and ruining human bodies. This means that irregular flooding, river pollution, undrinkable water, failed harvests, deaths of children in abandoned coal mines are because of coal extraction: they are a climate vocabulary.

Why does the vocabulary of people who live around extractive zone disappear from the negotiating table at COP 26 or at previous COP meetings? The answer is because mainstream, Western-biased knowledge divides society and nature. Climatology separates climate change indicators, such as carbon dioxide, as external to the community and more-than-human nature (Lohman, 2019). Effectively, climate change is separated from its cause, extractivism. The mainstream is keeping the climate change narrative away from everyday life because the climate crisis is considered a threat to capital accumulation that depends on extractivism. The mainstream solutions directed by the state, corporations, elites, and international NGOs are framed in technical, scientific language, and implemented on a massive scale: they expand the operation of extractivism.

The community in Samarinda shows how coal extraction in their territory, and its impact on families and communities, is an inseparable part of 'climate change'. Farmers' representatives in Makroman and residents of the city of Samarinda began discussing the relationship between village and city, rural and urban, coal mining, and climate change in 2012. They established a citizen movement called "Samarinda Menggugat". They used their climate vocabulary to sue the Indonesian government for failing to protect Samarinda citizens from coal extraction and its contribution to climate change. In 2013, the representatives of "Samarinda Menggugat" brought a lawsuit to the Samarinda District Court; it became the first citizen lawsuit in Indonesia (Toumbourou, 2014). One of their demands was to urge the Indonesian president and the East Kalimantan governor to close hundreds of abandoned coal mines in Samarinda. Their case was twice won in the city and provincial high courts in 2014 after 27 court trials but lost in the Supreme Court in 2016.

Samarinda Menggugat was connecting coal and climate—teaching us about climate vocabulary. Mai's account shows the ways in which climate change vocabularies emerge (when introduced by the state, corporations and NGOs from outside the community) and were submerged when the community centred its case around environmental justice.

Alice: I am so often hesitant to bring my experiences and observations of the onshore oil industry from the South of England into conversation with the testimonies Mai shares of the loss of lives, livelihoods and ways of life associated with open cast coal mining in Indonesia. I acknowledge the experiences of extractivism in England are incomparable to colonial contexts, yet there are commonalities in the way extractive logics and power are imposed. The violences experienced in colonial contexts have provoked insightful multi-dimensional analyses of extractivism; learning from (and taking care not to appropriate or extract) these perspectives and critical analyses can inform an understanding of extractive logics and power relations here in the centre of empire and fossil-fuelled industrial expansion.

Here—at the small (approximately two hectares) Horse Hill oil production site, set back from an oak-lined road between suburbia, Gatwick airport and privately-owned countryside—the experiences of extractivism are unspectacular. Perhaps the most pronounced way extractivism is evident is in the continued support given to the industry by the council, regulators, police and central government which overlooks and implicitly denies evidence of environmental and climate risks. Many of these risks are invisible, from the chemical fumes that sometimes surround the site, to the changing pressures and chemistry deep underground that pose a potential risk of seismicity and groundwater pollution, to the greenhouse gas emissions when the oil is combusted and the associated impacts of anthropogenic climate change. For the campaigners objecting to Horse Hill, the challenge is not only that these risks, which do not dramatically or directly affect local communities, can be difficult to mobilise around, but moreover that the systems of national planning policy and local governance are not designed in a way which accounts for the potential social and ecological costs of the proposed project.

Climate change does however present an opportunity through which the planning committee's decision can be challenged. As with the "Samarinda Menggugat" case and the plethora of climate litigation cases pursued by citizens and NGOs over the last decade, the law is being sharpened as a tool with which to fight polluting projects and, in turn, draw attention to the local impacts and injustices of extractivism. Following the local authority's decision to retain and extend the oil production site in 2019, local campaigners with the support of environmental NGOs have challenged the legality of this planning consent on climate grounds. The Judicial Review case centres on the failure of the council's Environmental Impact Assessment to take into consideration the climate change impacts resulting from the combustion of the estimated 3.3 million tonnes of produced oil. The potential of this case to have a national impact on planning policy and a global impact in terms of greenhouse gas emissions elevates Horse Hill from a local 'NIMBY' planning dispute to an emblematic struggle against fossil fuel extractivism as a root cause of the climate crisis.

Somewhat paradoxically, the legal challenge to the Horse Hill decision puts its faith in the systems, institutions and epistemologies upon which extractivism also relies. In the politically conservative area in which Horse Hill is located, the legal appeal is regarded as a respectable route for campaigners and has received significant financial support from locals. Before climate change hit the mainstream in 2019, thanks to the publicity brought to the issue by Extinction Rebellion and Fridays for Future, Horse Hill campaigners found their climate change concerns failed to engage the public or could be politically divisive. Meanwhile, the (intentional) confusion and lack of clarity surrounding onshore oil extraction techniques and their relation to the fracking industry had made it difficult to mobilise people around unknown but potentially significant risks. Peaceful anti-fracking protests and direct action at Horse Hill and at other sites in the area posed a threat to company operations, and the companies sought to deter protest by pursuing legal injunctions. At Horse Hill, the company was able to essentially buy a far-reaching injunction against 'persons unknown', preventing anyone from partaking in specified legal activities which could interfere with the profitability of the company. Campaigners successfully challenged this attack on their right to protest, and the injunction was scaled back as a result, but recently introduced laws continue to criminalise dissent by dramatically increasing the punishment and sentencing of peaceful protest. Whilst the state continues to reshape policy and law in the interests of corporate extractivism, the judicial review appeal seeks to flip accusations of criminality and remains the centre point of the campaign.

Putting climate change at the centre of the campaign and legal challenge makes Horse Hill emblematic of the UK government's willingness to sacrifice both the countryside and the climate to fossil fuel interests. Many Horse Hill campaigners care deeply about climate change, sometimes based on their own international experience in less economically developed (previously colonised) countries and an understanding of the global impacts and injustices of climate change. Others have become more recently concerned by climate change, with heightened awareness brought not only through activism but through lived experience of record-breaking heatwaves, droughts and energy prices which will impact the people and natural environments they love and care about. Connecting both local and global climate concerns to Horse Hill as a site of climate culpability, opens the opportunity to consider this local experience as part of a constellation of globally dispersed struggles against both the nearby and distant experiences of social, ecological and climatic impacts of extractivism.

We see this as an opening for building solidarity and for activism that addresses the ongoing coloniality of climate change by positioning citizens everywhere against extractivism everywhere, casting blame firmly on polluting and land-grabbing industries and the institutions they rely on rather than falling into the guilt-traps of individualistic (carbon footprint) or mis-anthropic ('humans are to blame') climate activism.

Alice: At a demonstration staged at Horse Hill as part of the Global Day of Action for Climate Justice during COP26, campaigners shared a recorded message from campaigners in Mozambique bringing a legal challenge against a UK government agency for funding a new mega gas project incompatible with the Paris Agreement climate commitments. The project has already forced thousands of people out of their homes and livelihoods and fuelled violent conflict and human rights abuses in the Cabo Delgado region. In the recording, the campaigner from Justiça Ambiental/ Friends of the Earth Mozambique expressed their solidarity with the Horse Hill case, urging those in Britain to support both cases and put pressure on the justice system.

Across the vastly different everyday experiences of extractivism, here is an opening through which the us/them narrative can be reimagined towards a common struggle against extractive corporations and the state systems that enable them.

How Is a Mainstream Climate Narrative Expanding and Deepening (the Coloniality) of Extractivism?

The previous dialogue illustrates the strategic mobilisation by activists of the climate change narrative as a way of connecting cause and effect, and as a (legal) tool with which to fight the local, everyday impacts of extractivism. In the same contexts, the climate narrative has also been used (co-opted) by extractive industries and states to legitimise the expansion and deepening of extractivism. Writing about the lithium triangle in Latin America, Voskoboynik and Andreucci (2022) describe how state and corporate discourses that justify extractive projects extend beyond an association with modernity and development, towards a strong ecological imaginary. Lithium extraction is presented as environmentally benign, through narratives of climate change, sustainability and the 'green economy'. Their discussion invites us to reflect on the situatedness of climate narratives: who mobilises a climate story and to what effect?

Mai: Feminist political ecology recognizes the multidimensionality of power relations among and between humans and in more-than-human relationships, leading to a global climate change crisis. However, the mainstream narrative of climate change separates society and nature and, accompanied by a sense of global urgency, centres on the technological fix and market solution or is limited to various earthly indicators of climate change. It makes the problem of climate change seem unrelated to people's everyday activities. Using the language becomes a political strategy to keep away the issue of climate change from everyday people's survival and resistance. One example is by associating the "climate" with "carbon", leading to the reason and answer of climate change being all that is related to "low carbon", including "reducing carbon emissions", or "low carbon development".

In Indonesia, one of these low-carbon developments is attached to the development of energy projects, including geothermal, because it has low-carbon emissions, including low sulphur dioxide and nitrogen oxides. One example of such projects is the Wae Sano geothermal project on Flores Island. In the local media, the geothermal company even asked residents and local government to play an active role in overcoming the increase in earth's temperature due to climate change by supporting the geothermal project. However, the community is worried that this project will dry up the biggest lake in the area, which is the source of agricultural irrigation and water for daily needs. On June 7, 2022, I met Yoseph Erwin, one of Wae Sano's residents who resisted the geothermal project. "One of our small sources of water, which was originally clear and hot and can be used to boil eggs, has now turned hot and yellow, even though it is only at the project exploration stage", he said. Wae Sano people who refuse the geothermal project are not only labelled as anti-development but also have the potential to be labelled as a climate unfriendly. Those who use a "carbon" narrative are not only narrowing the perspective used to understand the roots of climate change but are also potentially supporting misguided solidarity with all that is claimed as "low carbon", further supporting the green guise of oppression.

Alice: As the climate crisis worsens, a rapid and radical phasing out of fossil fuels is urgently required. This truth is obscured by policies and narratives that legitimise new fossil fuel extraction as part of the solution. The adoption and co-option of the language of climate action by industry and state to legitimise new oil extraction at Horse Hill in the South of England is an expansion and deepening of both the physical frontiers and the logics of extractivism. Following a decade of protest and diminishing public support for onshore hydrocarbon exploration in England promoted as an opportunity for energy security and economic growth, the mainstreaming of climate change concerns in 2019 provided a new vocabulary of 'climate mitigation opportunity' upon which onshore oil extraction could be pinned.

Rather than demonstrable facts, the climate mitigation opportunity narrative pushed by UKOG (the Horse Hill operators) is based on loose and generalised commitments to reducing carbon emissions which refuse to concede economic prosperity to socio-ecological viability. At the Surrey County Council planning meeting in 2019, UKOG representatives made a number of claims attesting to their commitments to emissions reductions and support for the energy transition, including the claim (found in court to be unevidenced) that oil produced at Horse Hill would have a lower carbon footprint because of energy savings from transportation. Acknowledgements of the need for decarbonisation were always caveated by the explicit assumptions that this should not risk "current levels of prosperity" and hence "oil and gas will have a significant role to play for some time to come". This closely followed the policies and language of Net Zero 2050 and Transition adopted by the government to legitimise the logics of incremental change and postponement rather than immediate climate action addressing the root causes.

By paying lip service to climate concerns and framing economic benefits as incontestable, the company sought to undermine the claims of climate activists and portray them as naive or ignorant. This was further exemplified in the planning meeting by UKOG's claims that they "are as committed to contributing to and safeguarding our local environment as any of our detractors", co-opting the narrative of environmental care in an attempt to add to their own credentials whilst casting the genuine environmental concerns of the public as irrelevant. This tactic was used more explicitly to greenwash and legitimise their case that new oil is needed to support a low-carbon economy, claiming "even Greta Thunberg endorses the use of oil-based products by sailing in a high-tech yacht made of strong lightweight oil-derived carbon fibre composite". This comment caused members of the public at the committee meeting to break their silence with expressions of disbelief; the audacity of claiming a renowned climate activist would support their project made campaigners feel insulted and gaslighted.

Through this combination of spin and mistruths reliant on the vocabulary of climate action, the greenwashing of oil production effectively denies the already unfolding reality (and coloniality) of climate impacts and delays climate action. New resource frontiers, from Indonesian islands to the English countryside, are legitimised as sacrifice zones to support the growth of a 'green economy'. As the physical frontiers expand, so too do the depths of extractivist logics and relations. The narrative of climate change as a call to slowly decarbonise rather than to rapidly degrow the economy legitimises the creation of new 'green' industries and requires new or rebranded 'green' extractivisms. That extractivism can be legitimised by the climate change narrative illustrates the need to push for narratives that foreground system change and target industries responsible for multiple intersecting socio-ecological injustices rather than 'just' the impacts of carbon emissions.

The Coloniality of Climate Responses

Our second set of dialogues reflects on how the meaning of 'green actions' originating in the Global North with the aim of reducing carbon emissions travel to the Global South. We ask, what effects do the travelling of these ideas, which may be presented as a form of global environmental care, have on the local communities in terms of their livelihoods and lives? How do these ideas travel and what are the risks involved in generalised narratives, solutions or solidarities that do not have a nuanced understanding of the different forms or modes of production? Here, Eunice and Dian reflect from the contexts of pastoralists raising livestock in Kenya and smallholder communities in Kalimantan, Indonesia who are seeking livelihoods from the cultivation of oil palm in the spaces between the large-scale corporate plantations.

Eunice: Since the launch of the United Nations Food and Agriculture Organizations (FAO) Livestock's Long Shadow report (Steinfeld et al., 2006) and more recently, the EAT-Lancet Report on healthy diets in the Anthropocene (Willett et al., 2019) calls to reduce consumption of livestock and livestock products as a solution to reducing greenhouse gases emissions have become common. There has been a massive dietary transition to vegetarian, vegan and white meat diets. While acknowledging that livestock production, like all other forms of agricultural production contribute to GHG emissions and must be aligned to mitigation efforts, these reports fail to differentiate between different ways of raising livestock, which have varying impacts on the environment (Scoones, 2021). Livestock and livestock products are blamed, rather than the method and scale of production. In addition, these mainstream climate narratives fail to recognize the benefits that some of the production systems like extensive pastoralism have for safeguarding the environment, livelihoods, and human lives. These benefits include reducing poverty, expanding livelihood opportunities, improving access to protein in diets, providing transport to the local communities (García-Dory et al., 2021). Thus, the focus ought to be on the production system rather than livestock and meat itself.

Pastoralism is an important form of extensive livestock production where pastoralists keep cattle, goats, sheep, camels, yaks, reindeers, and llamas on rangeland environments like deserts, savannas, steppes, arctic tundra, Mediterranean hills, and mountains, where alternative feasible livelihoods do not exist (Scoones, 2021). Such low input livestock production systems use rangelands with minimal inputs, have a lower climate, biodiversity, and water impact than the current climate narratives suggest. Compared to the intensive industrial systems, these extensive production systems can offer broader livelihood and ecosystem benefits. Through skilled grazing and different forms of mobility, pastoralists make use of pastures, grasslands, and shrubs, making the most of variability and climate related uncertainty (Scoones, 2021). Generalising all livestock production as harmful risks destroying low impact pastoralist livelihoods in Kenya that have nothing to do with damaging industrial livestock systems. This highlights the need to differentiate the impacts and contributions to environmental degradation between intensive and extensive livestock production.

To a large extent the narrative of a product, rather than its production method, being bad for the environment resonates with the oil palm case-explained by Dian, where the effects of palm oil boycotts are already being experienced by the local communities growing oil palm. Although the boycotts of livestock and their products have yet to take shape in Kenya, it may be a matter of time as such information travels across the globe. Already, several of my friends and peers have heeded to the rampant calls to boycott livestock and livestock products and have converted into vegetarianism and veganism for environmental reasons. This trend will be fast tracked by the strong presence of European expatriate community in the country who are already searching for vegetarian options in the local restaurants, thus increasing demand. While their commitment is admirable, I often feel there is a gap in understanding what exactly makes livestock bad for the environment. The narrative risks boycotting products from people who have contributed minimally to the current global crisis we are going through, thus punishing them further.

The intrinsic connection between the Maasai and their cattle is stronger than just financial gains. Although there is the financial benefit of livestock keeping, most of them consider their livestock as important members of their family, illustrated in how they care for and relate with their livestock. Unlike in my community and others engaging in intensive livestock production, where the calf is separated from its mother upon birth, the Maasai cow and calf remain together until a new calf is born. This means that the calf continually suckles until it's mature enough. I also noticed that the calves were allowed to suckle on one side as women milked the cows. This allows the mother–calf relationship to blossom. Livestock is often counted according to their parental lineage rather than numbers. Every evening the household women would count the livestock in relation to their mother's. It was also common to purchase livestock to be slaughtered in the markets rather than slaughter one of their own. One study participant explained that his familial relationship with his livestock deterred him from slaughtering them: "It is like slaughtering a family member". Many development NGOs find this paradoxical, especially during droughts, where livestock owners risk going without food rather than slaughter one of their animals. As an example, during my fieldwork, one of the study participant's cow's udder was eaten by hyenas at night. When the herders reported the incident to the owner, I assumed that the cow would be slaughtered for meat immediately as its capacity to produce milk had been compromised. To my surprise, the owner sought a traditional healer's services to sew the udder and apply medical plants to ensure it healed. The owner knew the cow wouldn't produce milk but still held on to it rather than sell it off to the butcher. Upon inquiry, she responded that it was her best cow and had given her a lot of milk in its lifetime, and she would continue taking care of it even though it may never produce milk anymore.

Dian: Eunice reminds me of the supposedly green action of boycotting palm oil for its association with deforestation. Both actions, which find purchase initially with consumers in Global North, purport to care for the more-than-human. Yet the next in the queue—rural people in the Global South—are lost when these actions travel over space to the land of producers.

In the Global North, consumers perceived the action to boycott oil palm as the only way to save the planet. When I was grocery shopping in Brighton in the UK, I read "this product does not contain oil palm" on many items—this lack of palm oil was framed as an intrinsically good quality and a selling point. There was constant news in the mass media telling me that oil palm cultivation is a main contributor to climate change and the culprit behind the forest fires, orangutan killings, and deforestation.

In the oil palm community context, where a previously forested landscape was replaced by large-scale oil palm, everyday life changed in ways that varied across communities, depending on intersections of power based around gender, class, ethnicity, migrant status, proximity to local power, and peoples' relation with oil palm. The oil palm companies also matter. While researchers agree that large-scale oil palm companies negatively impact the landscape, humans and more-than-humans, there are some differences in experiences: where a few companies left communities with some wriggle-room for survival and where other companies did not. Where communities have this wriggle-room, some small-scale farmers have been inspired to plant the tree themselves and benefit from it. While mostly we hear about large-scale corporations and their extractivist impacts, in Indonesia (which produces most palm oil in the world), 40% of total oil palm area is accounted for by small-scale independent farmers. For these smallholders, planting oil palm trees is to improve their livelihood and to care for their family, not for accumulating profit at any cost to the humans involved, as in large-scale companies. The goal is survival and bettering life for future generations. Below, I draw out the different stories that come from these communities, based on research I undertook whilst working as a research officer at CIFOR (Center for International Forestry Research) in Indonesia.

An indigenous middle-aged woman who comes from a lower economic background showed us her everyday life. She wakes at 3am to start caring for her rubber plot, and to cook and prepare her 6-year-old child for school before boarding the truck that takes her to the oil palm plantation where she is employed as a casual worker. She earns less than 6 USD per day and works from 7am to 3 pm. She occasionally needs to tend her rice field after her work in oil palm plantations. She said that what she earns is barely enough to get by every day and meet her family's needs. She worries that her first daughter, who is in high school, might not finish school. She hopes that her two daughters have a better life than hers (CIFOR, 2017a).

A second story comes from a young Indigenous couple who are permanent workers on an oil palm plantation, receiving a monthly salary, with extra if they harvest more than the target. They said that achieving the target is not easy but doable if they work hard. The couple have a small child who is taken care of by the couple's parents when the couple works in the plantation. For this couple (the husband is a migrant from another village), oil palm gives them hope to start their own plots and a small shop for their future. They said that they don't want to let go of their harvester's work for the company even if they have their own plot and small shop in the future already (CIFOR, 2017b).

A final story comes from an Indigenous leader in the village who has managed to save some land. He has started his own oil palm but is anxious as he has no access to knowledge and necessary resources (seedlings, fertilizer, pesticides, etc.) to do it properly. He strongly states that if the companies can benefit from oil palm, the villagers should be able to as well. He witnessed other communities who have prospered from oil palm. He anxiously waited for his two-year-old oil palm trees to show results. Oil palm trees begin to fruit after three years—if the fruit is bad at this point, then the trees are bad trees, and they have to be cut. The first important step to plant oil palm is making sure that the seedlings are good. But access to this information is difficult if you don't know who to ask. As he puts it: "If those companies can make a lot of money and improve their life from oil palm in our land, why should we only watch? While we were here from the start?" (CIFOR, 2017c).

As we reflect on these stories together, we note the forms of coloniality that re-emerge when green actions in the Global North are taken without careful regard to the nuances of everyday lives in communities in the Global South that are themselves under threat from extractivism. Superficial understandings of community experiences mean green consumer actions originating in the Global North risk extending injustices when communities get swept up in broad-brush actions, and where political actions are not targeted at the extractivist systems that are doing harm. Specifically, what this can mean is a foreclosure of more sustainable, reciprocal ways of relating to animals through the ecosystems of pastoralism, and to the land and forest, through smallholder oil palm cultivation that presents possibilities for replenishing rather than depleting lives and landscapes.

Extractivism's Other: Concluding Reflections on Alternatives

Mai

While the durian falls or is picked,
transported by wooden boats
brought to the house, neighbours, and local market

When the 'tanah air' is extracted,
transported by iron barges,
taken somewhere unknown
(In between spaces, Despite Extractivism exhibition 2022)

I imagine the durian, a forest tree that reaches 30–50 m high, with its thorny fruit skin and soft, fragrant flesh of the fruit. Kalimantan Island is the centre of Durian biodiversity. Biologists found 30 species of durian grow in Kalimantan, with various skin fruit colours, some are yellow, called Lay fruit, and Keruntungan is red. Durian tree mark out land tenure in Indigenous communities, telling the story of a family's lineage and connection to land. Durian fruit is also a source of cash for education fees (Maimunah & Agustiorini, 2021).

In Sungai Lalang, Central Kalimantan, the durian season is a joy, marking the arrival of the fruit season. Durian trees are planted along with other fruit crops scattered in people's yards, tree-gardens and forests along the river. The aroma of the fruit invites wild animals such as wild boars, Mawat (fruit bats), binturong (weasels), and various types of birds and nectar-eating beetles to approach. For the Murung people, this is the time to hunt—while waiting to harvest the swidden. Men and some women hunt pigs and other animals in the forest in the group. The durian season means the season of collectivity. The activity in the village can move to the durian forest until harvest time arrives. I saw small boats full of durian fruit going back and forth on the Lalang river in the afternoon and evening.

Consuming and processing durian fruit also requires communality. We can eat fresh durian or consume it after it is processed into lempok and tempoyak. Lempok is a durian lunkhead that can be stored for a long time, while tempoyak is fermented durian flesh. In fermentation, microbes break down the sugar and fat compounds to produce a healthier food.

Fermentation reduces the harmful effects of durian and diminishes harm to our bodies, others, and the world around us (Fournier, 2020).

The durian season is a sign of inter-species relations, in contrast to the extractivist relations that govern mining of coal in Kalimantan since the colonial period-relations that harm and extract from old forests, rubber plantations, fields, and orchards, including durian trees. Mining removes topsoil, revealing solid black rock with a strong odour and combustion smell. The black rock is taken and transported via hauling roads before finally being sent on the rivers out of Kalimantan Island.

Alice

On a paved road, under the flight path of Gatwick airport, on the edges of suburbia, in the heart of empire, outside an oil extraction site, is it possible to imagine alternatives to extractivism? An alternative to fossil fuel extraction is to 'leave it in the ground' and to instead pursue renewable energy sources, but as we have seen it is often not enough to replace extractivism with green extractivism—the resources required for renewable infrastructure and the corporate nature of the industry often come with environmental injustices. Perhaps less pragmatic but more critical are the alternatives to the logics of extractivism, an abusive and dominating way of seeing the Other as a resource to serve goals of accumulation.

In a handful of small but intentional ways, those opposed to oil production at Horse Hill have thought and practised together some alternatives to extractivism. By staging protests, picnics and ceremonies at the gates of the site, activists not only draw attention and bear witness to local and climate impacts of the operations but also subvert extractivist logics by physically occupying the space with our own sets of logics. Poignantly, the Faith at the Gate events involve the sharing of readings, reflections and silent meditation or prayer. These are occasions to celebrate the abundance of nature and observe the changing seasons and to stand in solidarity with others fighting climate, environmental and social injustices here and elsewhere. Through these expressions of reverence and care, a powerful juxtaposition is staged between the peaceful gathering at the gates and the sacrilege of the disregard for the Other being perpetrated on the other side of the gates.

Near the Horse Hill site are deposits of clay, historically used in the area for brickmaking. Local potter Xanthe Maggs was inspired to find ways to use this clay to support the Horse Hill campaign, such as through community outreach workshops and the creation of ceramic badges.

Joining in with these experiments with clay, I was struck by the question of what distinguishes our extraction of this material from the extraction of the oil, and this in turn opened an invitation to consider what other, more reciprocal kinds of relationship could be had with this land. Extractivism is not only about the scale or the consequences of the extraction, but also about the intent: extracting a resource for personal gain is inherently distinct from extracting a material for the purpose of creatively inspiring care for the land and climate. Clay creations have been auctioned to fundraise for the legal case, and a clay bead travelled from Horse Hill to Glasgow with an activist joining the 'Camino to COP26' pilgrimage, walking across the UK to bring messages from communities to the COP.

Care for the land through walking and being in the landscape seems to be capturing the popular imagination in England at a time of increased recognition of the physical and mental health benefits of being in nature that emerged from the Covid-19 lockdowns, and the increased awareness of nature's vulnerability as the climate breaks down. New campaigns for the Right to Roam, including days of peaceful Mass Trespass, have been supported by some Horse Hill campaigners and draw attention to the lack of public access to the English countryside and the controversial history of private land ownership and inheritance by elites. This includes land acquired by slave owners through the publicly funded compensation they received as a result of the abolition of slavery. Access to nature in England remains intimately tied to colonialism in such ways, and the campaign seeks to encourage responsible access to nature to counter disconnection and exclusion from the land. This campaign goes hand in hand with campaigns against the industrialisation of the countryside through extractive projects and has the potential to bring care for local nature and land rights into conversation with care for global climate impacts, injustices and extractivisms.

We began this chapter with our reflections from COP26, refracted through our different positionalities and rooted networks within the coloniality of contrasting extractive contexts around the world. As we have woven the threads of our stories together, our dialogues have been knotted around the root causes of climate change—extractivism, injustice and disconnection from nature. When we draw the coloniality of extractivism more closely into the weave, we see how extractive injustices are being reproduced in green economy 'false solutions' and are perpetuated in broad-brush green actions emanating from the Global North, where insufficient regard is paid to nuanced community perspectives. Our reflections and stories connect with the efforts of activist movements to decolonise climate and environmental justice and to mount a robust challenge to the simplifications that arise from fixating on carbon emissions without addressing systemic issues that derange human and more-than-human relationships on and with land and water. Feminist political ecology provides us with the tools to create a closer weave, threading through an analysis of extractivism with lived experience, of communities and of ourselves as researchers and activists. Through FPE, we attend to situated knowledges that shape storytelling in all its forms, from the climate vocabularies of corporate and state actors to the languages mobilised to tell stories of extractive harms.

We have closed our chapter with reflections on the possibilities for what Sultana (2022) evocatively refers to as the restructuring of relationships to ecologies, waters, lands and communities to which we are intimately, materially and politically connected. As we juxtapose our reflections and stories, we listen for ways to recover or amplify sustainable alternatives to the logics of extractivism. In closing the chapter, we open up the possibilities within FPE, which provides a convening space for exploring the opposite of extractivism: relationships between humans and more-than-human natures based around stewardship, reciprocity, regeneration and ensuring life for future generations through healthier ways of relating to the land.

Funding: This chapter was funded by the Wellbeing Ecology Gender and cOmmunities Innovation Training Network (WEGO-ITN) funded by the European Union's Horizon 2020 research and innovation programme under the Marie Sklodowska-Curie grant agreement No. 764908-WEGO 2018-2021.

References

Aykut, S. C., Pavenstädt, C. N., Datchoua-Tirvaudey, A., D'Amico, E., Braun, M., Hinks, E. K., Schenuit, F., Wilkens, J., & Rödder, S. (2022). *Circles of global climate governance. Power, performance and contestation at the UN Climate Conference COP26 in Glasgow* (CSS Working Paper Series No 4). Universitat Hamburg Center for Sustainable Society Research. https://doi.org/10.25592/css-wp-004

Bhattacharyya, G. (2018). *Rethinking racial capitalism: Questions of reproduction and survival*. Rowman & Littlefield.

Chagnon, C. W., Durante, F., Gills, B., Hagolani-Albov, S., Hokkanen, S. M., Kangasluoma, S., Konttinen, H., Kröger, M., LaFleur, W., Ollinaho, O., & Vuola, M. P. S. (2022). From extractivism to global extractivism: The evolution of an organising concept. *The Journal of Peasant Studies, 49*(4), 760–792.

Chao, S., & Enari, D. (2021). decolonising climate change: A call for beyond-human imaginaries and knowledge generation. *eTropic: Electronic Journal of Studies in the Tropics, 20*(2), 32–54.

Chua, L., & Idrus, R. (2022). Introduction: Unpacking indigeneity in Southeast Asia. *Sojourn: Journal of Social Issues in Southeast Asia, 37*(1), 1–26.

CIFOR. (2017a). *Gender and palm oil: A day in the life of a female palm oil worker* [Video]. YouTube. https://www.youtube.com/watch?v=ilw_6r-PVug

CIFOR. (2017b). *Gender and palm oil: Working together as a couple* [Video]. YouTube. https://www.youtube.com/watch?v=XjuOuqeVafU

CIFOR. (2017c). *Gender and palm oil: Staying independent* [Video]. YouTube. https://www.youtube.com/watch?v=DhCHjzVzA00&t=65s

Fournier, L. (2020). Fermenting feminism as methodology and metaphor: Approaching transnational feminist practices through microbial transformation. *Environmental Humanities, 12*(1), 88–112.

García-Dory, F., Houzer, E., & Scoones, I. (2021). Livestock and climate justice: Challenging mainstream policy narratives. *IDS Bulletin*, Online First. https://doi.org/10.19088/1968-2021.128

Gómez-Barris, M. (2017). *The extractive zone: Social ecologies and decolonial perspectives*. Duke University Press.

Haraway, D. (2019). It matters what stories tell stories; it matters whose stories tell stories. *A/b: Auto/Biography Studies, 34*(3), 565–575.

IPCC. (2022). *Climate Change 2022: Impacts, adaptation, and vulnerability.* Contribution of Working Group II to the Sixth Assessment Report of the Intergovernmental Panel on Climate Change (H.-O. Pörtner, D. C. Roberts, M. Tignor, E. S. Poloczanska, K. Mintenbeck, A. Alegría, M. Craig, S. Langsdorf, S. Löschke, V. Möller, A. Okem, & B. Rama, Eds.). Cambridge University Press.

Li, T. M., & Semedi, P. (2021). *Plantation life: Corporate occupation in Indonesia's Oil Palm Zone*. Duke University Press.

Lohmann, L. (2019). *Climatology/ideology* (World Rainforest Movement Bulletin, No. 247).

Maimunah, S., & Agustiorini, S. (2021, July 19). Durian and the Coloniality of Power (Part I). *Sudostasien*. https://suedostasien.net/durian-und-die-kolonialitaet-der-macht-teil-i/

Malm, A. (2016). *Fossil capital: The rise of steam power and the roots of global warming*. Verso Books.

Merchant, C. (1980). *The death of nature: Women, ecology, and the scientific revolution*. HarperCollins.

Moore, J. W. (2015). *Capitalism in the web of life: Ecology and the accumulation of capital*. Verso Books.

Pereira, C., & Tsikata, D. (2021). Extractivism, resistance, alternatives. *Feminist Africa, 2*(1), 1–13.

Scoones, I. (2021). Pastoralists and peasants: Perspectives on agrarian change. *The Journal of Peasant Studies, 48*(1), 1–47.

Sultana, F. (2022). The unbearable heaviness of climate coloniality. *Political Geography*, 102638.

Sundberg, J. (2016). Feminist political ecology. In D. Richardson, N. Castree, M. F. Goodchild, A. Kobayashi, W. Liu, & R. A. Marston (Eds.), *International encyclopedia of geography: People, the earth, environment and technology* (pp. 1–12). Wiley.

Shahbanu, A., & Maimunah, S. (Eds.). (2018). *Extractive oligarchy and the declining quality of people's life. How the power of coal extraction oligarchy in Indonesia makes the state neglectful toward the declining quality of residents' life*. Deadly Coal Series, Volume 4, Jaringan Advokasi Tambang (Mining Advocacy Network). Retrieved June 9, 2022 from https://www.jatam.org/wp-content/uploads/2018/12/deadly-coal-series-english.pdf

Steinfeld, H., Gerber, P., Wassenaar, T., Castel, V., Rosales, M., & de Haan, C. (2006). *Livestock's long shadow: Environmental issues and options*. Food and Agriculture Organization of the United Nations.

Tokar, B. (2020). Introduction: Climate justice and community renewal. In B. Tokar & T. Gilbertson, *Climate justice and community renewal: Resistance and grassroots solutions* (pp. 1–16). Routledge.

Toumbourou, T. (2014). *Indonesian Lawsuit Pushes Local Government to Regulate Massive Coal Mining Industry*. The Asia Foundation. Retrieved August 30, 2022, from https://asiafoundation.org/2014/10/15/indonesian-lawsuit-pushes-local-government-to-regulate-massive-coal-mining-industry/

Veltmeyer, H., & Petras, J. (2014). Introduction. In H. Veltmeyer & J. Petras (Eds.), *The new extractivism: A post-neoliberal development model or imperialism of the twenty-first century?* (pp. 1–20). Zed Books.

Voskoboynik, D. M., & Andreucci, D. (2022). Greening extractivism: Environmental discourses and resource governance in the 'Lithium Triangle.' *Environment and Planning E: Nature and Space, 5*(2), 787–809.

Willett, W., Rockström, J., Loken, B., Springmann, M., Lang, T., Vermeulen, S., Garnett, T., Tilman, D., DeClerck, F., Wood, A., Jonell, M., Clark, M., Gordon, L. J., Fanzo, J., Hawkes, C., Zurayk, R., Rivera, J. A., De Vries, W., Sibanda, L. M., ... Murray, C. J. L., W. (2019). Food in the Anthropocene: The EAT–Lancet Commission on healthy diets from sustainable food systems. *The Lancet, 393*(10170), 447–492.

Willow, A. J. (2018). *Understanding ExtrACTIVISM: Culture and power in natural resource disputes*. Routledge.
Wolford, W. (2021). The Plantationocene: A Lusotropical contribution to the theory. *Annals of the American Association of Geographers, 111*(6), 1622–1639.

CHAPTER 3

Extracting Us: Co-curating Creative Responses to Extractivism Through a Feminist Political Ecology Praxis

The Extracting Us Curatorial Collective

Alice Owen, Siti Maimunah, Dian Ekowati, Rebecca Elmhirst, and Elona M. Hoover

Introduction

Extractivism—the logic that enables large-scale resource extraction and the exploitation of people and nature—has inspired a raft of responses from artists and creatives. Often, these responses take shape through

A. Owen (✉) · D. Ekowati · R. Elmhirst · E. M. Hoover
Centre for Spatial, Environmental and Cultural Politics, University of Brighton, Brighton, UK
e-mail: A.Owen@brighton.ac.uk

D. Ekowati
e-mail: Dee.ekowati@gmail.com

R. Elmhirst
e-mail: R.J.Elmhirst@brighton.ac.uk

W. Harcourt et al. (eds.), *Contours of Feminist Political Ecology*, Gender, Development and Social Change,
https://doi.org/10.1007/978-3-031-20928-4_3

aesthetics of the industrial sublime, with vast and person-less landscapes of ruination which provoke an awe-inspired inertia. Such portrayals, while effectively conveying the spectacular scale of resource extraction (Koch, 2022; Traynor, 2021), tend to overlook the everyday violence of extractivism in the lives and life worlds of those living in spaces that colonial logics and extractive capital have constituted as 'extractive zones' (Gómez-Barris, 2017) or who are otherwise entangled in the web of extractivism. Moreover, the agency and ongoing-ness of communities and individuals living despite or in resistance to extractivism is often hidden from view.

This chapter shares the journey of a curatorial collective that takes on the challenges of looking at extractivism through the lens of feminist political ecology (FPE). It tells the story of an initiative called Extracting Us which was born out of a meeting between scholar-activists based in the United Kingdom and Indonesia; of conversations, meetings and the sharing of worlds and experiences. From this encounter came a desire to speak to issues of extractivism very specific to a geographical context (East Kalimantan), but that we knew could cross boundaries, borders and times. Engaging with FPE, we foregrounded the experiences and agency of communities on the frontlines of this extractive zone, paying attention to how power over the environment—which follows the contours of coloniality, race and patriarchy—not only reshapes landscapes but limits the possibilities for thriving, with devastating consequences. With this came a need to shift perspectives of extractivism from the universalising, the spectacular, and extractive aesthetic objectification towards curating creative forms from the frontlines of extractivism. These perspectives instead centre the gaze on the everyday as the time–space where extractivism is both experienced and resisted.

In the three years that our curatorial collective has worked together, 'extractivism' as a theme and 'organising concept' (Chagnon et al., 2022) has become much more prominent in academic, activist and artistic discourse and practice, exceeding its earlier definitions as a political-economic regime of commodity extraction and export (Acosta, 2013) to consider extractivism as a mindset or way of thinking (Willow, 2018).

S. Maimunah
University of Passau, Passau, Germany
e-mail: Mai.jebing@gmail.com

Black and Indigenous perspectives on environmental justice highlight colonialism and racial capitalism within extractivist logics (Chagnon et al., 2022; LaDuke & Cowen, 2020; Wynter, 2003). We therefore situate Extracting Us in a wider set of conversations that were evolving and unfolding along a similar path to us (for example, Arts Catalyst, 2019; Tsing et al., 2017; Ureta & Flores, 2022, The Global Extractivisms and Alternatives Initiative (EXALT) at the University of Helsinki). Extracting Us has grown collectively, from 2019 to the time of writing in 2022, through our curation of a series of three exhibitions and accompanying events with artists, activists and researchers engaged in what Gómez-Barris (2021, p. 855) refers to as 'the arts of land and water defence'. The process of collaborative curation, with its emphasis on creative-political enquiry and on thinking and doing 'otherwise', has also led us to think anew (and across our different positionalities) about what we understand as extractivism and how this is informed by the practice of FPE.

We weave together our conversations about the curatorial process, exploring different ways of articulating and responding to extractivism through integrating art, activism and research. In doing so, we speculate about the ways in which these influence each other. Our process for this chapter involves us each reflecting upon the various points of our journey through the curation of Extracting Us at ONCA Gallery in 2019, Extracting Us online in 2020, and Despite Extractivism online in 2021–2022. In recording our reflections, we adopt an FPE practice that coalesces around plural ways of seeing, thinking and learning relationally. The text therefore retains messiness as we shift between collective and individual voices. We seek to avoid flattening out our differences within our collaborative curation: a long process involving three exhibitions, dozens of institutions and hundreds of people across different continents.

We have organised our reflections around four elements of the project: the process of curating and building our collaboration across boundaries, the practice of weaving materials and conversations through the exhibitions, the care-full interventions we have made as we join up art, academia and activism in extractive contexts. and finally, the ways this project has expanded our understanding of extractivism and our practice of feminist political ecology.

Curating Across Boundaries

It all started with the story around the Mahakam River, the second largest river in Indonesia, located in East Kalimantan, Indonesia. Here (there), the largest coal exploitation site in Indonesia is slowly eroding life around the city of Samarinda. One of us, Siti Maimunah (Mai) has embodied experiences working with JATAM (Jaringan Advokasi Tambang/Mining Advocacy Network, Indonesia) and a group of mothers whose children drowned after falling into abandoned mining pits in East Kalimantan a decade ago. From her encounters around the Mahakam River, the story flowed through meetings with Elona Hoover, Becky Elmhirst, Dian Ekowati and Alice Owen in Brighton, and then the idea of the exhibition emerged.

The confluence that brought stories of extractivism from the Mahakam river into conversation with researchers and activists in Brighton, UK, was enabled by the EU-funded Wellbeing, Ecology, Gender and cOmmunity (WEGO) early stage researcher network that brought two of us (Mai and Dian) from Indonesia to the UK to an initial collaboration with Alice and Becky (WEGO early stage researcher and mentor respectively) and post-graduate researcher Elona Hoover in Brighton in the summer of 2019. Brighton is a city on England's south coast, far from East Kalimantan, but linked through the hidden threads of UK corporate investments that profit from mineral exploitation. It is also a place where conversations and actions in search of ecological justice are flourishing, in spaces such as ONCA Gallery and the University of Brighton's Centre for Spatial, Environmental and Cultural Politics (C-SECP).

ONCA Gallery is a Brighton-based arts charity 'that bridges social and environmental justice issues with creativity' (ONCA, 2022) and was also an important locus for young people in Brighton who were engaged in the #FridaysForFuture Youth Strikes for Climate. Conversations that took place between Mai and Elona in ONCA's kitchen and backyard coalesced around the everyday impacts of coal extraction in Kalimantan, and serendipitously led to the first exhibition titled '*Extracting Us. Looking Differently: Feminism, Politics and Coal Extraction*'. The name of 'Extracting Us' emerged from the philosophy of the Mollo indigenous peoples in Indonesia: 'destroying nature is like destroying our human body'.

The planning, discussions and activities that swirled around this exhibition brought the five of us together—Elona, Mai, Alice, Dian and

Becky—as a curatorial collective: we are the 'We' in this chapter, each bringing our own experiences and ideas. Yet, this 'We' also risks conflating individual positionalities and standpoints: different experiences of living and researching in Global North and South contexts, different life/work stages, varying care responsibilities in differing everyday time-spaces, and different postures, skills and knowledges in relation to research, activism and art practice.

This 'We' also falls short of embracing all those that are part of this story. The first exhibition (Brighton, July 2019) involved Mai's activist networks, notably JATAM, who helped with design work and communication for the exhibition, and the London Mining Network. WEGO provided a relational space of feminist political ecology learning and funding for many of our activities, C-SECP provided material and intellectual support for engaged research activities connecting creativity and environmental justice, and the ONCA Gallery directors provided practical and critical input and connections with activists and further artists-researchers.

We began to feel our way... The collective continues to be shaped by the collaborations that grew along the way and that helped us establish our curatorial principles, including challenging North–South narratives on extractivism, foregrounding accessibility, and in bringing in the perspectives of those most affected. These collaborations also enabled us to curate across boundaries—of knowledges, geographies and temporalities.

Elona: I feel grateful for the human connections, for the relationships we made through the process. For the way in which it allowed me to express my own and each of our potentials: artistic, academic, activist, and practical. We could use knowledges and capacities that are often boxed into different spheres.

Becky: I felt engaged, stretched, challenged and at times, outside my comfort zone. Some aspects, particularly early on, made me feel deeply uneasy—a version of imposter syndrome perhaps, pushing against disciplinary boundaries that I wouldn't normally traverse, and entering into ethical territory that I perhaps had avoided previously.

It felt like we formalised the curatorial collective when we proposed a second iteration of the exhibition for the POLLEN (Political Ecology Network) conference—originally to be hosted in Brighton in June 2020 and moved online due to the pandemic to September 2020. This involved making our curatorial principles more explicit as we invited contributors to join an expanded 'Extracting Us' project that sought to bring together feminist political ecology perspectives and extractivism. These principles articulated feminist political ecology loosely, inviting contributions that thought about extractivism in terms of materials from (and of) the earth as well as in terms of human and more-than-human experiences and energies. We sought contributions that listened to perspectives from those most affected, and that included narratives of resistance.

Elona: I was excited about being able to create and maintain connections through the pandemic, finding ways of adapting the exhibition and modes of collaboration with curators and contributors in troubled times... while giving space to those for whom adapting was not possible or even violent.

The attention and care for contributors also extended to our collective. This became particularly important when we crossed the boundary from physical to virtual, shifting the collaborative Extracting Us exhibition online in the (northern) Spring of 2020 because of the COVID-19 pandemic. As we dispersed to different 'homes' and countries, we held our collective together with regular meetings, working across four time zones (UK, France, Germany and Indonesia) and checking in on each other. Countless hours of discussions, tensions, laughs and tears. Collaborating and trying our best not to extract from each other. Some individuals still work more than others. Some feel regret that they cannot contribute more. Some lost good friends, fell ill, took care of ill loved ones. Yet our growing familiarity with being together online in the strange intimacy of each other's home spaces helped dissolve physical distances, even as our experiences with the pandemic diverged. Slowly and organically, we took time to consider, reflect, and think about how We produced knowledge and work related to our daily lives and in relation with others.

Dian: The Extracting Us curatorial collective helped me through the pandemic and the loneliness during our PhD study and its solitude. We had good news when a member of the collective (Elona) was expecting and delivered a beautiful person in 2021.

Alice: Moving online actually made for different kinds of connections. In those early days of the pandemic there was a sense of connecting despite and because of our isolation. Catching up with and working with the collective heightened my awareness of the privileges I experienced in my daily life (health, home, job and economic security, loving family and friends) and gave me a stronger awareness of the time involved in care responsibilities and commitments that I did not have. Similarly, the curation process allowed me to recognise material differences between the context shared in the exhibition and the context through which I have been researching extractivism—oil extraction with relatively minor local impacts in a relatively privileged part of the South of England. This awareness often led to feelings of guilt, and to a kind of paralysis from not knowing what to do with these privileges. It made me feel like I should be doing more for this project, yet at the same time I didn't want to impose and risk losing the horizontality and collectivity.

Becky: Although connecting online for meetings and webinars was a balm early on in the pandemic, we quickly became aware of highly differentiated and uneven experiences of COVID-19 pandemic as this overlaid and amplified existing inequalities associated with coloniality and extractivist logics. Farhana Sultana has written powerfully about the pandemic, climate and coloniality as overlapping crises (Sultana, 2021) and clearly each of us was differently positioned in relation to this. When we met with contributors online, personal introductions were punctuated with harrowing stories of bereavement and fear in the face of failing health services or inexplicable political responses in different corners of the world. Reciprocity and care via Zoom became an integral part of our feminist practice.

Elona: After we had launched the end of the second iteration, I also had a clear desire to pass things on. I had been leading more at that moment but did not want the process to be tied to one person. I think of this as a feminist practice of attachment and detachment (Ahmed, 2012). It was then my turn to take a step back, and with the third and last iteration I felt overwhelmed when I was not able to contribute any more—trusting that I would not be left out when I found the time again.

Our reflections narrate a messy process, as we worked with the boundaries of each other's realities, tried to learn to relate and sought out ways to reflect that in our work. Donna Haraway's (2016) idea of 'staying with the trouble' describes our process well: it reminds us of the layers of difficulties and emotions faced by those who live in an extractive landscape, and how we continue to care for each other. We also connect with Cindy Katz's description of 'messiness' in trying to constitute 'the field' where this is not something that can be sanitised from the researcher's everyday life and positionality (Katz, 1994). The story of how we sought to curate across boundaries (of experience, geography and coloniality) and the journey of our collaboration amidst the structural pressures of academia, activisms and global pandemic is something akin to a river. Like the flow of a river, it was shaped by a palimpsest landscape and, encountering unexpected obstructions, was disrupted and diverted without diminishing its essential qualities.

Weaving Materials and Conversations

Our second set of reflections converges around what we learned as we worked with the creative materials of the exhibitions (the photographs that comprised the first exhibition in 2019, the works submitted by contributors to the online Extracting Us exhibition in 2020 and its later iteration in 2021, Despite Extractivism). The process of curating exhibitions and events involved working with the collected materials to unravel conversations, draw unexpected connections and pursue the lines of inquiry that these inspired.

Our first exhibition, curated with the ONCA Gallery team in Brighton, staged 46 photographs from a larger collection gathered over the years by Mai and her collaborators in Indonesia. Mai and Elona spent hours carefully selecting the photographs, initially trying to create a contrast between visions of destroyed or empty landscapes and photos of everyday lives. In creating juxtapositions, the exhibition aimed to show how families and communities are continually devastated by the impacts of coal mining and abandoned coal pits, which have a colonial legacy linking to the UK where the exhibition was shown.

Alice: For me the moment when I really 'got it' was when we all gathered to group the photographs. It was the first time I'd properly seen them, and the stories of the people in the photographs which Mai recounted to us were suddenly so visceral and powerful. In that room of women, mothers and daughters, I sensed the gravity of our responsibility to share and make sense of the stories of the women and children who looked back at us from the photographs. In that session, experimenting with ways of grouping and juxtaposing the images, I began to see the importance of a feminist political ecology lens for extractivism. For me, it's about paying attention to and taking seriously the ways conflicts over nature are experienced through the everyday and through embodied and emotional registers. This applies both 'in the field' as we come to understand the experiences of the communities we research with, and in ourselves as we think and feel our way through our research practices in ways not detached from the rest of our lives or selves.

In the gallery, the photographs were arranged in five clusters. Each centred around a large-scale photograph which showed extractive landscapes with massive physical damage, such as floods surrounding the entire Samarinda city, the loss of farmland, abandoned toxic pits, landslides, and the impacts on villages. These were juxtaposed with smaller photographs of dead children's faces in coal pits, uncovering the predatory character of the mines: extractivisms destroy and deprive social relations, womb relations, knowledge and care. Photographs of resistance actions were included in every photograph cluster, capturing the agency of communities in responding while avoiding pathos and victim narratives.

The emotional and personal connections evoked by the photography and the accompanying invitations to respond were imagined as an embodied experience that could foster response-ability (Haraway, 2016). This goes beyond the typical academic work of exemplifying, adding complexity to, or communicating the impacts of extractivism. Cultivating response-ability involves creating a sense of connection and of agency rather than being struck by the inertia of guilt or the overwhelming scale of global extractivism.

Elona: Though I felt like an 'outsider' in terms of feminist political ecology, I felt like I was able to bring a different perspective on how to express some of the theoretical issues in practice: what affective strength different images had together, the importance of the aesthetic, ways of putting images on walls, or designing postcards and imagining together how to involve people affectively in an exhibition. This was striking for me when I created the overlay of the children's faces with the mining pit for the exhibition poster. Twenty faces. A deep pit. The scale of which I start to see as the massive but minuscule mining machine hidden in its bowels came to my attention. Dark earth. Lost lives.

Alice: Installing the exhibition at ONCA was a strange experience, enjoying being together in the gallery space and getting excited about the exhibition whilst confronting again the harrowing images. I remember in particular how we positioned the photographs of Rahmawati, a mother who had lost her child, at eye level—it gives me goosebumps even now recalling the intimacy of those face-to-face encounters.

Dian: Despite coming from Indonesia, I did not 'get' it before, even when Mai talked deeply about these issues. The stories of Raihan and other lives lost in the mining pits are not something that I read in everyday mainstream media. I remember sitting with the photos and the collective in a small room in our university and how depressed I was by the lost lives and mothers losing children, and by my own ignorance of this situation (Fig. 3.1).

Maybe it was the quality of the photos, taken with phones, or photos of photos taken in school, or photos of family portraits. There is an intimacy to the non-professional. These images brought home the everyday violence of extractivism in ways images of vast extracted landscapes found in a political ecology textbook cannot.

Moving the second Extracting Us exhibition online was a challenge for the curatorial collective and contributors. We had to move from a predefined exhibition space to an undefined online space, where we wanted to create a feeling of visiting somewhere, of being immersed in the work of different contributors, as well as including human interactions and shared learning in a virtual space.

Fourteen contributors responded to our call for works relating to feminist political ecologies of extractivism for the second Extracting Us exhibition, sharing work from Tajikistan, Ireland, Trinidad and Tobago, Brazil, Eastern Himalayas, Indonesia, Zambia, UK, Ecuador, Senegal, France and India. Several had planned an object or performance-based

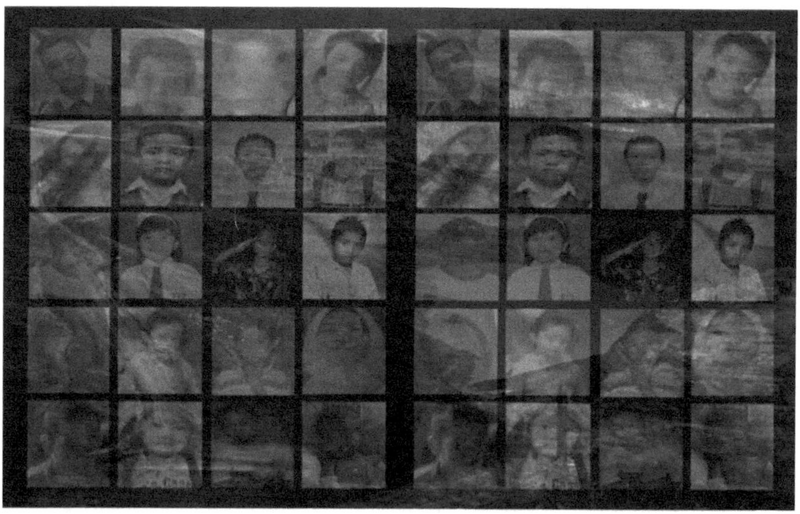

Fig. 3.1 Open cast coal mine, overlain with images of children who drowned in abandoned coal pits (East Kalimantan, Indonesia)

contribution and found it challenging to move towards the digital. Conversations focused on how to work through that without losing an essence of the work and if it was worth it. For instance, Sandro Simon's work (focused on Senegal) was a series of six video loops that were meant to be played on their own and/or simultaneously. Another contributor, Maica Gugolati, whose work focuses on Trinidad, explained how she had to first imagine an overcrowded soundscape in an embodied physical space, and after that, it made sense to watch it online.

Our third exhibition, Despite Extractivism, was also online. In this iteration, we brought together contributions from Indonesia, UK, Ireland, Sweden, Senegal, the Urals in Russia and Spain, this time exploring how communities care, resist and persist despite extractivism. For this, we worked with independent curator Celina Loh, whose professional insights on curation and the online exhibition development encouraged us to consider questions around how we were communicating, with whom and why. She helped us create a more 'tactile' virtual experience.

In both online exhibitions—Extracting Us and Despite Extractivism—we had accompanying events where participants were also asked to engage with physical objects in their surroundings. For example, we held a webinar on creative engagements on the front line presented with exhibition contributors at the POLLEN20 conference in which we invited participants to respond creatively to what they were listening to and seeing (Fig. 3.2).

We noted an affective involvement by asking people to draw and inviting them to be engaged while they listened. We shared the doodles towards the end of the webinar, including participants in the conversation without necessarily sharing words. This active embodied engagement—in the context of the ongoing pandemic and online webinar saturation—became something we wanted to build on as we gained more confidence in working through online spaces. As well as artists' and activists' presentations and discussions, the webinar series accompanying Despite Extractivism invited audiences to engage in a different register by listening to songs, poems, guided reflections and movement.

For example, during Arabel Lebrusan's 'Toxic Waves' contribution to Despite Extractivism at our webinar on embodiment, participants were invited to simultaneously draw waves in time to 270 beats of a metronome, representing the 270 people killed in the collapse of the Minas Gerais iron ore tailings dam in Brazil in 2019. Through this performance, Arabel explored the questions 'Can art making, through

Fig. 3.2 Sharing creative responses to the Extracting Us webinar at the online POLLEN conference (2020)

embodied thinking, activate our empathy at a deeper and more instinctive level than our rational understanding of events? Can this urge us to act? Can this help us grieve?'

Dian: The body movement of making the wave as part of Arabel's participatory performance art made me feel emotional. I felt anger, grief, stress, connection with the community displaced by extractive forces. I felt very tired afterwards (the event was held quite late from where I joined), but when I tried to sleep on my comfortable clean bed and breathing clean air, I remembered those whose lives have been extracted through mining and do not have a clean bed to sleep on (and no clean air to breath in) in the nights after struggling during their days. There were feelings of solidarity and of guilt for not being able to give more 'real support'.

Care-Full Interventions

Integrating relational practices such as the work with materials mentioned above was part of our curatorial process. The notion of 'care' emerged as a major tributary to our river through our embodied experiences of a feminist curation process and through the themes of our work together and independently. In this section, we reflect on the ways we worked through a politics of connection and solidarity, aligned with our curatorial principle to develop solidarity actions aligned with the narratives of resistance that were highlighted in the contribution of the three exhibitions.

As part of the first exhibition held at ONCA, we developed a series of postcards that people could write and send to a range of people/actors relevant to the context of coal mining in Indonesia as featured in the exhibition. This included the mothers who had lost their children, local NGOs, and local and national Indonesian politicians. As well as being in the gallery for people to write independently, postcard-writing was integrated into the event series. The exhibition was accompanied by a film screening, talk, and workshop with youth climate protestors involved in #FridaysForFuture. Ibu Rahima's story was featured in the exhibition, and exhibition visitors wrote and sent her solidarity postcards (Fig. 3.3).

Ibu Rahima's 14-year-old son Raihan drowned in one of the abandoned coal pits in East Kalimantan. At the time when Indonesians were celebrating Mother's Day in 2014, Ibu Rahima demanded that the Government close and clean up more than 250 abandoned coal pits in the city of Samarinda. She visited Raihan's former school, spoke to the students about staying away from the pits and collected 10,000 signatures for a petition which she gave to the Indonesian Minister for Environment and Forestry. It was important to us to avoid pathos and a victim narrative in the exhibition: as her story shows, Rahima is not a passive victim, she is a survivor.

One year after the first exhibition, we received a short video from Rahima. She was holding the postcards that had been sent by the exhibition visitors. The exhibition had helped to uncover the connection between Indonesia and the UK; the latter had benefited from the first coal extracted and shipped from East Kalimantan for use in fuelling colonial trade and warships. In her video, Rahima addressed the exhibition

Fig. 3.3 Solidarity postcards ready to be sent from the UK to Indonesia (2019)

participants: 'Thank you for friends from the UK who have supported me. With the blessings of God Almighty, I will continue the struggle to get justice for the children who drowned in the (abandoned) coal pits'.

Becky: At first, many of the contributions in the first exhibition and Extracting Us online exhibition gave me a profound sadness—a hopelessness, or rather, a sense of my own hopelessness in the face of what communities were enduring and responding to. Was visibilising the 'unseenness' of extractivism's violence sufficient? Taking part in the postcard writing activity (and learning about the replies that came back from the community), shifted my perspective as I recognised the possibilities for relating otherwise through co-learning and solidarity across the fault lines of racial capitalism and the coloniality of extractivism.

The process and reflections directed us to further develop curatorial principles based on care-full interventions that: (i) challenge 'north–south' narratives on extractivism, listen to perspectives from those most affected, and develop actions of solidarity and resistance across countries and continents; (ii) include narratives of resistance where possible/relevant; and thus avoid relying on pathos that might develop an 'us/them' feeling; and (iii) develop solidarity actions during the exhibition, for instance engaging emotionally and physically with the exhibition material.

These principles formed part of our invitation to contributors, and all were invited to consider how opportunities for these kinds of solidarities could be integrated into their contributions. Every contribution included website links to further information and to specific groups they could support, or actions visitors could take. As curators, it is hard for us to know how effective this has been in terms of those on the frontlines of extractivism being or feeling supported. For us, what has been more apparent is the way we have begun to build connections between the contributors, the communities we are all connected to through our work, and our visitors.

We began weaving connections between artists for the online exhibition, asking them for instance to visit the exhibition and join and video call to comment on each other's work. As the curatorial collective, we had already spent some time finding our connections (the themes of the exhibition), but different ideas such as 'time' emerged as potent concepts in these discussions and the conversation flowed in unexpected directions. Many artists found exciting resonances between others' work and their own and others, whether in terms of the content or the artistic modes and practices engaged with. When we took these conversations into the more public space of the POLLEN webinar, other kinds of connections were made between the contributions and those participating.

Building on the richness of these discussions and the engagement we felt from the online participants, for the Despite Extractivism iteration of the exhibition we more deliberately curated the series of online time spaces. Something we pay attention to in feminist political ecology is how systems of power operate and are resisted across scales, and we decided to theme the Despite Extractivism exhibition event series around this idea, starting with the body and embodiment, expanding to community, and finally considering worlding. These events took careful planning, inviting artists, researchers and community groups to connect in advance and prepare presentations or provocations. The stories told through the exhibition and discussion workshops opened up new ways of understanding care as creative expressions which enable us to better recognise and analyse multiple and often hidden ways of tending to each other and more-than-human natures against or despite extractivism.

For those involved, there were poignant moments of connectivity, despite our physical distance. During one of the Despite Extractivism events, where contributor Dewi Candraningrum (Indonesia) discussed her paintings of the Kartini Kendeng ecological defenders, she was accompanied by the background sound of a cricket chirping the onset of the dry season in Central Java. This reminded us that the gossamer threads of solidarity that connected us across cyberspace involved other species beyond our own.

Our efforts to convene events that would enable us to feel, experience and share together across worlds were not always successful: the material limitations of online connections could let us down. A particularly regretful moment came when the Kartini Kendang were scheduled to share their resistance song, but uneven bandwidths reminded us once more about the ways uneven infrastructures replicated wider global injustices and inequalities, and taught us once more about the importance of accessibility if our feminist response to extractivism activism (or, following Willow, 2018, our extrACTIVISM) was to have any real purchase.

Not all the contributors were able to join our events. In part, this was because of the 'flowing' nature of our work, evolving slowly rather than according to a master plan. We also lacked the capacity to make the events truly inclusive by using alternative text for images or by ensuring accessibility for those with hearing disabilities.

Becky: Our feminism aims to be postcolonial and intersectional—I'm not sure I can quite describe it as decolonial. Some of our aspirations in this direction have been difficult to fulfil—moving online meant dealing with various forms of digital exclusion, and we have always struggled with how to address the issues that come from working predominantly in the English language, even as in the later work we sought to include simultaneous translation in some of our events.

Expanding Orientations

Like the flow of a river, the story and bodily experience of creating and curating the Extracting Us exhibitions has branched in unexpected directions. It doesn't matter whether we have flowed with the main river or along one of its branches. Originally the Extracting Us exhibition was about mining coal and the predatory relationships and violent logics of extractivism. The story of extractivism along the Mahakam River is connected with extractivism in rivers, mountains, cities, provinces, countries and other continents, and with the people in them. In the exhibitions that followed, rather than following a particular commodity, our course followed the undulations of the landscape as we continued to explore these connections with an expanded group of contributors and collaborators.

As we expanded our orientations, our reflections and inquiries have followed and contributed to the evolving contours of feminist political ecology as a way of thinking through and resisting global socio-ecological injustices. Yet it also feels like the more we have collectively learned, the harder it is to communicate and maintain a space for learning that is open across all backgrounds, capacities, positionalities and ontologies. It can be easy to forget what it was like to not have even heard of 'extractivism' or 'feminist political ecology', and we have become aware that we have developed our own vocabulary between us over the years, which perhaps makes it more difficult to communicate in everyday language or ideas. Reflecting together, we see how the questions and messages of the exhibitions have evolved, but connecting these into something that feels like a stable 'mode of inquiry' can be challenging.

Alice: I think we (I) felt somewhat reluctant to offer up our own definitions as we wanted to keep alive the sense of collective inquiry with our contributors, communities and audiences. Yet as curators we have a responsibility to offer up useful explanations and guiding ideas.

The feminist impetus to avoid controlling, channelling and containing the flow of the narrative has always been important, and this means that while some of our learning is collective, it is also embodied and situated, sedimented in various ways among us. Here, we share some learnings from the process of making unexpected connections between the works that feature in the exhibition series.

Mai: My own contribution 'Between the frontier spaces' has many resonances with 'Between the rivers' by Daniel Macmillen Voskoboynik, both reflecting on rivers. Daniel describes his family home in Russia, and the kindness of the river, which changes gradually as the landscape is impacted by extractivism. 'We lost our footing to the earth, but we never knew it,' he writes. Daniel describes one of the characteristics of extractivism that makes what was originally meaningful and sacred become cheap and meaningless:

> This vision—also known as extractivism—enforces a misunderstanding. Life, which is sacred, is actually cheap, and valuable only in its service to the economy. What matters and what being means are not ongoing questions, but simple equations.

Similarly, I find a contradiction in my work between a photo of rusty tools left by a logging company in heavily damaged and deforested land now being mined by coal companies, and the continuation of rubber and fruit orchards providing economic, social and ecological benefits. The contradiction I describe in the form of a poem:

> When the forest, no longer with trees,
> Abandoned wreckage machine
> and the logging company left
>
> While the orchard is evergreen,
> Stocking for harvest every season,
> and the rubber dripping sap every day

Like Daniel, I understand that extractivism is not a sectoral project but a perspective and way of operating: with a large-scale, predatory character. The predation of extractivism is short, making the sustainable and the sacred worthless, no longer usable or requiring a long time to heal. Extractivism is in stark contrast to daily activities such as gardening, planting, and performing rituals that become the daily routine of Murung Indigenous people, the face of 'Despite Extractivism'.

Becky: I had an academic understanding of extractivism and approached this from a feminist political ecology framing, which emphasised research that was designed around words: texts and conversations, albeit those trying to centre perspectives from communities, and focusing on everyday livelihoods. I had spent much of my FPE life in ethnographic fieldwork, aiming to deploy a feminist sensitivity. Engaging with artists and creatives, I learned the value of seeing the different registers through which communities tell their stories, and through which others tell the stories of communities, through art, creativity, song, performance. What was striking for me was how specific and yet universal some of these practices were: from the connections to poetry, song and place-making in the Sperrin Mountains in the North of Ireland (V'cenza Cirefice) to the ways in which indigenous Waoroni women chanted their histories of connection to their home territories, in the face of, and despite violent extractivisms in the Ecuadorian Amazon (Margherita Scazza). Extracting Us coincided with an amplification of decolonial and anti-colonial thought and action, and with the centring of 'extractivism' as a concept in political ecology. The Extracting Us collective provided a space in which to explore the connections between coloniality and extractivism.

Alice: The exhibitions and events helped me understand the pervasiveness of the violent logics of extractivism, be this through large scale devastation of landscapes and the direct loss of life and livelihoods, or through the onto-logics of extractivism which deny other ways of being in and relating to the world.

Working on the website or the Extracting Us online exhibition required me to spend time with the works in a deep and focused way. This was really influential and inspirational to some of the work I presented, not to mention ongoing thinking about extractivism. I was particularly influenced by V'cenza Cirefice's video overlay showing the submerged perspective of a river flowing through the Sperrin Mountains in the North of Ireland, and the same area represented from the disembodied perspective of a map which abstracts living, flowing connections and, through the 'Extractive Gaze' (Gómez-Barris, 2017) sees the land as a resource.

One idea we discussed while curating the exhibition was how extractivism(s) are not all pervasive. Sometimes defining something can make it seem bigger than it is, but it is also important to explore how there are many alternatives to extractivism already in existence. As we explored in our events and conversations, solidarity and resistance can also occur across multiple, intersecting scales.

Elona: Being part of the collective introduced me to the notion of extractivism and allowed me to develop more connections with feminist political ecology scholars and thinking. It allowed me to also see my work on urban commoning in a different way: making a link between the extractive processes of capitalism in relation to urban space and practices of enclosure.

Dian: The curation process shifted my understanding from avoiding being extractive, towards being care-full. Extractivism was to be avoided not only in terms of extractive methodologies in our research work, but also in practices. I was convinced not to do fieldwork during the pandemic as the field would be the oil palm community residing in places with little to no access to healthcare.

Mai: As a collective that grew up in the FPE space, we have also grown and become more sensitive to understanding extractivism, which transfers not only material but also the resistance to it. Resistance exists in forms not often recognised as resistance, such as in the unseen work done mostly by women of collecting water, gardening, keeping rituals before the harvest or meeting with mothers affected by the coal mine. This care work is extracted under the operation of mining, logging and other extractive projects. We are starting to learn something new, something about spaces of care as a way to look at and understand resistance in the extractive landscape.

Becky: The shift towards Despite Extractivism felt like a point where the collective moved from wielding a hatchet to sowing seeds (to use a political ecology metaphor) or at least, tending what was emerging through the cracks, where I felt our learning moved towards a focus on extraACTIVISMS (Willow, 2018). This has carried over into other areas of my work, principally my political ecology pedagogy where different knowledges and prefigurative politics now feature. Now, I feel an ethical imperative not only to point to what's wrong, but to share and learn where alternatives (not false solutions) are being practised, however small and mundane, and to see everyday creative practice as a valuable methodology.

We have 'looked differently' at extractivism from many angles, in many ways and at different scales, thinking alongside creative interventions of various kinds. This has helped us recognise how extraction/extractivism has uneven impacts, and to appreciate what is at the core of 'extractivism': the coloniality of its logics.

Moving Forward—Weaving, Thinking, Caring, Acting

From differently positioned researchers with shared interests to a curatorial collective, the *Extracting Us* journey has been at once an experiment in feminist political ecology as praxis, and in 'the exhibition' as an iterative space for co-inquiry and public engagement on the less explored or otherwise unconnected aspects and contexts of extractivism.

We have journeyed together with each other, the many contributors, and the communities and individuals who have shared their experiences along the way, all giving and gleaning differently from this project and coming to pause on different riverbanks rather than simply going with the flow. We have found spaces to stay with our troubles, from the personal to planetary, with Extracting Us providing something of an anchor in ever-more turbulent times. Our motivations, understandings, and positionalities within structures of power and privilege will always situate us differently in relation to extractivism, but we share a sense that it is a privilege to have had the opportunity and the collective energy to curate these exhibitions.

As our river flows towards the sea, we are confronted with multiple ways in which extractivism in its many registers is re-inscribed, its logics becoming more pervasive. Extractivism intersects with the rising tides of patriarchy, authoritarianism, violence, racism and oppression. How to continue this project of understanding extractivism in a way which weaves connections between contexts, allows for care-full response-abilities towards active solidarity and expands the 'we' in ways that do not flatten difference will certainly be troubles we will have to stay with. Whether through academic, artistic or activist work, or work that continues to erode the boundaries between these, more creativity will be needed.

Creativity, communities and care have orientated our co-curation process, from the ways we work together to the ways we work towards amplifying, connecting and learning with communities impacted by extractivism and the artists, academics and researchers who work with them. Through these experiences and the stories shared in the exhibition contributions, our thoughts converge on the idea that extractivism is a violent and pervasive way of enacting force that dismantles human and more-than-human communities and the relationships of care that exist within them. Yet. Is this the end of the story? How do communities—be these communities of place, communities of practice or communities of solidarity—continue to find different registers through which to question, subvert, resist, persist and care? How best can we continue to create and curate creative and care-full spaces across boundaries?

Funding: This chapter was funded by the Wellbeing Ecology Gender and cOmmunities Innovation Training Network (WEGO-ITN) funded by the European Union's Horizon 2020 research and innovation programme under the Marie Sklodowska-Curie grant agreement No. 764908-WEGO 2018-2021.

References

Acosta, A. (2013). Extractivism and neoextractivism: Two sides of the same curse. *Beyond Development: Alternative Visions from Latin America, 1*, 61–86.

Ahmed, S. (2012). *The cultural politics of emotion*. Routledge.

Arts Catalyst. (2019). *Assembly—Extractable matters*. Retrieved 20 June 2022 from https://www.artscatalyst.org/assembly-extractable-matters

Chagnon, C. W., Durante, F., Gills, B. K., Hagolani-Albov, S. E., Hokkanen, S., Kangasluoma, S. M., Konttinen, H., Kröger, M., LaFleur, W., Ollinaho, O., & Vuola, M. P. (2022). From extractivism to global extractivism: The evolution of an organizing concept. *The Journal of Peasant Studies*, advance online publication 9 May 2022. https://doi.org/10.1080/03066150.2022.2069015

Gómez-Barris, M. (2017). *The extractive zone: Social ecologies and decolonial perspectives*. Duke University Press.

Gómez-Barris, M. (2021). Submerged perspectives: The arts of land and water defense. *Globalizations, 18*(6), 854–863.

Haraway, D. (2016). *Staying with the trouble: Making Kin in the Chthulucene*. Duke University Press.

Katz, C. (1994). Playing the field: Questions of fieldwork in geography. *Professional Geographer, 46*(1), 67–72.

Koch, P. (2022). *Extraction: Art on the edge of the Abyss*. Retrieved 22 June 2022 from https://www.extractionart.org/home

LaDuke, W., & Cowen, D. (2020). Beyond Wiindigo Infrastructure. *South Atlantic Quarterly, 119*(2), 243–268.

ONCA. (2022). *ONCA gallery: About Us*. Retrieved 8 November 2022 from https://www.onca.org.uk/about-us

Sultana, F. (2021). Climate change, COVID-19, and the co-production of injustices: A feminist reading of overlapping crises. *Social & Cultural Geography, 22*(4), 447–460.

Traynor, M. (2021). *Extractivism: Preview of A Glorious Ruckus*. Retrieved 22 June 2022 from https://www.extractionart.org/earthday

Tsing, A. L., Bubandt, N., Gan, E., & Swanson, H. A. (Eds.). (2017). *Arts of living on a damaged planet: Ghosts and monsters of the Anthropocene*. University of Minnesota Press.

Ureta, S., & Flores, P. (2022). *Worlds of gray and green: mineral extraction as ecological practice*. University of California Press.

Willow, A. J. (2018). *Understanding extrACTIVISM: Culture and power in natural resource disputes*. Routledge.

Wynter, S. (2003). Unsettling the coloniality of being/power/truth/freedom: Towards the human, after man, its overrepresentation—An argument. *CR: The New Centennial Review, 3*(3), 257–337.

CHAPTER 4

Ouch! Eew! Blech! A Trialogue on Porous Technologies, Places and Embodiments

Ilenia Iengo, Panagiota Kotsila, and Ingrid L. Nelson

Introduction

What kind of narratives and experiences sit at the margins of feminist political ecology (FPE) analyses of health, embodiment and environment? Ilenia, Panagiota and Ingrid started their trialogue as scholars and activists from different backgrounds and relations interested in this question. In this chapter, the stories we share are as much part of our fieldwork notes

I. Iengo (✉) · P. Kotsila
Institute of Environmental Science and Technology (ICTA), Universitat Autònoma de Barcelona (UAB), Barcelona, Spain
e-mail: ilenia.iengo@uab.cat

P. Kotsila
e-mail: panagiota.kotsila@uab.cat

I. L. Nelson
Department of Geography and Geosciences, University of Vermont, Burlington, VT, USA
e-mail: ilnelson@uvm.edu

W. Harcourt et al. (eds.), *Contours of Feminist Political Ecology*, Gender, Development and Social Change,
https://doi.org/10.1007/978-3-031-20928-4_4

as part of our personal diaries. These experiences transform how we do research, at the same time transforming us. What ties them together is our attention to the relationship between embodiment and environment. We come together thus to discuss the politics of making and remaking health in relation to socio-environmental processes, through the concept of porosity and the realm of technologies. For us, feminist political ecology is a convening space for thinking with the trouble in discourses and practices around health: from the biomedical sciences' gaze and the relational ecology between bodies and places, vis-a-vis the biopolitics of the nation-state to the far from innocent but complex role of both medical and digital technologies in navigating illness.

Recognising that there is ample feminist academic and activist work actively engaging with the concept of embodiment and with the gendered politics of health and medicine, we are particularly interested in how such work can infuse current thinking and activism on social and environmental health and technologies. We are not driven by a "gap" in literature, but rather by our own situatedness in and response-ability towards the embodied experiences of chronic pain, infection and nausea, and how these can inform non-normative experiences and knowledges surrounding health. We borrow the term response-ability from Donna Haraway, as in "proposing together something unanticipated" (Haraway, 2016, p. 130), starting from experiences, theories and epistemologies rarely depicted in the context of FPE and health.

We understand bodies as ecologies, not separated by what happens "outside our skin", but as the interconnected system of life within and beyond our bodies and the economic, racial, gendered, biophysical places we inhabit (Guthman, 2011; Guthman & Mansfield, 2013). Understanding embodied ecologies as relational allows us to see the boundary marked by the skin as connecting, instead of separating, our bodies and our environments (Iengo, 2022; Jackson & Neely, 2015). In this chapter we will reflect on "porosity", as the relational ecology between bodies and environments, departing from the ambivalent experiences of such relations, mediated by medical and digital technologies, as well as gender, race, and disability. We draw together theory and insights from critical disability studies, science, technology and society studies (STS), environmental justice, and political ecologies of the body and of health, to produce powerful elaborations that help queer and crip the binary division between bodies and places, against "the fantasy of self-contained wholeness" (Sadler, 2017). This does not negate the politics of bodily autonomy, but rather complexifies them by amplifying the demands for self-determination and reparation by BIPOC, working-class people, women, LGBTQIA+ and other marginalised subjectivities in response to the literal embodiments of differentiated and historically contextualised oppressions.[1]

Eco-crip theories constitute a fertile space from which FPE may learn to cast a light on the non-innocent acts and technologies of "seeing" which affect what counts as nature and natural (Jaquette Ray & Sibara, 2017). Such seeing questions and reinvents the ideologies of pureness and desirability for an "elsewhere" and "out there" to be controlled and tamed. And this reinvention influences power-laden notions of desirability and purity within human communities. In the encounter between feminist technoscience and crip studies, we learn about the nuanced and ambivalent entanglements of health, embodiment and technologies (Bailey & Peoples, 2017; Kafer, 2013; Mingus, 2010).

We weave together three stories from Italy, Greece and the United States to point to the tensions that emerge when environmental ethics and discourses do not take into account the intersectional experiences and positions of vulnerability that are often located at the deeply political, intimate and embodied levels.

Ilenia brings in her *chronic pain* experience as a catalyst to learn about how environmental injustice is accumulating in bodies and territories. In her southern Italian hometown, environmental injustice is inescapable. Grassroots socio-environmental justice activism helped many to politicise territorial experiences of illness and contamination, to learn the power of bodies coming together to halt slow violence (Nixon, 2011). In her quest for environmental justice Ilenia is influenced by the epistemic activism of crip communities in producing counter-knowledge on contested illnesses and mutual aid practices.

Panagiota tells the story of the embodied and invisible injustices suffered in romanticised agricultural production of "healthy and local" food products, through the exploitation, disenfranchisement and exposure of immigrant farmworkers to *infection* and ill health. As a white European academic returning to the olive grove-filled countryside where she used to spend her summers, the critical analysis of socio-environmental conflict and inequalities suddenly "hit home". She

[1] See Michelle Murphy's important intervention in these discussions in her 2017 article, 'What Can't a Body Do'.

observed the patterns of change and inequality in local agriculture, social fabric, labour conditions and health outcomes in the context of neoliberal austerity-ridden Greece.

Ingrid speaks of the ambivalent position of pregnant people depending on two technologies: big pharma-produced medication used in order to be able to deal with *nausea* and navigate place, and digital social media platforms as key spaces of solidarity and empowerment in the struggle to navigate disability and illness. She connects these technologies to the pedagogical spaces and institutional labour context in Vermont, with their attendant privileges and inequities.

Bringing together three distinct and distant stories, we point to how disembodied environmental politics can become exclusive and replicate injustice by showing the commonalities that emerge from embodied and situated perspectives of engaging with ecologies, places, technologies and health experiences.

This chapter brought us together as FPE scholars involved in the WEGO network and engaged in health, environmental justice and embodiment from different perspectives and angles, to engage in a process of knowledge production which happened via multiple digital encounters we held over two years, much of which was during the global COVID-19 pandemic. We started by sharing three stories that relate to these themes, and stem from our positionalities as scholars and activists in FPE. The process moved to a phase of "trialogue", inspired by Costa et al. (2015) that came out of a set of questions we asked each other in an effort to trace common and uncommon threads in our stories. What we share is a leitmotif of bodily ecologies of health, as well as the knowledge and discourses around them, being shaped both politically, through power, and culturally, through meaning.

Bodily Ecologies and the Embodied Politics of Visibility

Chronic Pain

In the summer of 2018, after 20 years of normalised pain and medical gaslighting, I received the diagnosis of endometriosis. I was officially on holiday, but the reality was that I spent days curled up in bed unable to move or talk because of the pulling, stabbing and twisting pain across my

> abdomen, lower back, and legs. As it happened cyclically, it was time to search again online for a gynaecologist and hope to get to the source of my pain. The ultrasound showed an endometrioma, an endometriosis cyst encapsulating the right ovary, strangely unseen until then. The gynaecologist explained there was nothing to worry about, instead there were plenty of options ahead, especially one. He turned to my partner and with camaraderie and a condescending tone exclaimed: ***"I recommend you go home and engage in 'you know what I mean'"*** openly advising sexual intercourse and pregnancy as a pleasurable (for my male partner) cure (for my gendered body).
>
> —Ilenia Iengo, personal notes

Ilenia: What time is it? How long have I been here? I often ask myself while tuning in and out from sleep, without rest, gazing at the walls of my bedroom. There is no linearity in time when a body experiences endometriosis, adenomyosis and fibromyalgia. Some call this coexisting, expanding, retracting, embodied, relational and reclaimed temporality "crip time". Alison Kafer helps us reflect on the queer genealogies of crip time not only as an accommodation for disabled bodies and minds but as "a challenge to normative and normalizing expectations of pace and scheduling" (Kafer, 2013, p. 27). Chronic inflammation is the common thread connecting the multiple illnesses that persist within my body and the symptom of the past, present and intergenerational damage occurring to the land. I grew up in the shadow of Mount Vesuvius, a geologically fiery territory nowadays called by local activists "The Land of Fires" due to the burning and spewing of toxic waste in dumps and landfills, in the southern Italian region of Campania. A few years ago, I discovered I have deep infiltrating endometriosis, a chronic inflammatory disease embodying the slow violence of Italian capitalist development and biocide (Iengo, 2022). During gestation and growth, toxic substances like dioxins and PCBs have soaked my territory in Southern Italy and crossed the porous boundary between the inside and the outside, they enveloped the bowels of my body in adhesions and endometriosis lesions fuelling the ever-growing familiar pain (Iengo, 2022).

Starting from the experience of chronic illness, where the intersection of gender, environmental, class and racial violence produce a complex system of oppression, I present some instances of situated strategies to reclaim our spaces, knowledge, voices and desires as disabled and chronically ill people in socio-environmental justice movements, while working towards cripping and queering the political relationship between health and environment.

The contaminated body is the terrain for subjectification and politicisation of everyday life (Iengo & Armiero, 2017). Environmental injustice and exploitation accumulate in marginalised bodies which are contaminated twice: first by the toxic substances occupying their neighbourhoods and lives, second by the toxic narratives which silence this injustice, invisibilise the sick bodies, and blame those who live with the consequences of contamination (Barca, 2014). Our communities and territories are considered expendable for the profit and well-being of others; therefore, we fall ill, reflecting the poor health of our neighbourhoods and cities. We get sick and, contrary to the dominant narrative, it is not our fault, nor the consequences of our lifestyles and choices. Some of us may live with chronic illness and disability as a consequence of disabling social and economic environments (Hedva, 2018). The lived experience of disability or chronic illness is confronted with the ableism of a system that classifies us as unworthy and unproductive. We are discarded at every turn of life due to the societally compulsory able-bodiedness and able-mindedness (Kafer, 2003; McRuer, 2002). Such narratives of obliteration, undesirability and aberration to be fixed are inherent even in the prefigurations of radical socio-environmental futures, where chronically ill and disabled bodies are often not contemplated, reproducing our absence (Piepzna-Samarasinha, 2018). The disability justice movement overturns the narrative: disabled and chronically ill people take the floor and do the fundamental work of constructing imaginaries and practices of socio-environmental indispensability (Pellow, 2018) to dismantle ableist ideas according to which chronic illness and disability are merely medical conditions while the material evidence of social and environmental injustice is erased (Kim, 2017). In this intervention, I recollect some of the thoughts, worries, feelings and developments of an ecotransfeminist exploration and expansion of a cripistemology of pain (Patsavas, 2014) together with the emergence of disabled and chronically ill folks' epistemic activism wherein we speak for ourselves (Hamraie, 2012).

I often wonder what would have become of this experience if I didn't have the lens of transfeminism to help me recognise that there was nothing normal or acceptable in being belittled when I came to the ER unable to move or speak out of pain, in the physician's suggestions of pregnancy as cure, or in the 22 years of diagnostic delay. Diagnosis is a fundamental tool in the path of chronic illness and is often linked to conflicting feelings of validation and discouragement. These feelings express the contradictory, yet impossible to disentangle, tension between the pathologisation that comes with diagnosis and the legitimation of pain to access support systems. This objective tool of biomedical science, a non-neutral field of knowledge with its eugenic, ableist, racist, misogynist, transphobic and industrial-military-complex roots, holds immense power to shape how the world sees and categorises us and our chaotic embodiment (Clare, 2009; Jaquette Ray & Sibara, 2017; Kafer, 2013; Lorde, 1980).

In recent years, with the composite transfeminist community in Italy, we are elaborating intersectional perspectives on health to include the gendered experiences of disabled and chronically ill folks. Through the feminist practice of self-awareness, we have learned to recognise the structural violence of sexism and patriarchy. Similarly, self-awareness among chronically ill and disabled people helps us to highlight how ableism is at the basis of capitalist, productivist and patriarchal society imposing the standard of productivity upon us that we ourselves have often metabolised. In the autumn of 2021, with the transfeminist network Non una di meno (NUDM) we began to reflect on our embodied experiences of patriarchal medicine and the multiple "invisible" and gendered illnesses. I prefer to use the term invisibilised, because even if some illnesses do not necessarily leave visible marks on the body, if one pays close attention to these lives, it all becomes very evident. On October 23, 2021, under the slogan "sensible-invisibile", we took our aching bodies to public space in about 20 Italian cities to share the experiences of people with endometriosis, adenomyosis, vulvodynia, fibromyalgia and pudendal neuropathy. We have organised online national and city-wide assemblies for the construction of a *presidio* to amplify the voices of those who live with these diseases, denouncing the misrecognition and disinterest of the national health care system. In the spirit of self-organised knowledge production, we produced and distributed leaflets with self-help tips, basic information concerning the different illnesses and banners explaining that experiencing pain is not normal and should not be normalised (Jones, 2016). Gendered illnesses reflect social taboos inasmuch as medicine reflects patriarchal oppressions, leading to the social, economic, psychological, health and environmental consequences borne individually by the sick person. In Naples, we gathered in a square with chairs and benches for everyone to rest and participate, where we shared our stories, hung posters where we imprinted the medical violence we have suffered, the desires that keep us going and reclaimed our experiences beyond the tragic and pitiful narrative, instead politicising disability and chronic illness as the complex and powerful experience, a place of knowledge production to be collectivised.

For the first time in the feminist movement in Italy, we are expanding the historical framework focused on reproductive health and abortion to include the needs and desires of our chronically ill and disabled bodies (Price, 2015). To express the anger that inflames our transfeminist political action is a form of healing from the accumulated trauma and violence (Hedva, 2018). The path for an ecotransfeminism that recognises and fights against ableism as one of the forms of oppression of white and colonial supremacist capitalism is a battle where no one is excluded and where the issues of mutual care, anti-sexism, interdependence, anti-racism, sex work, environmental and social justice can produce powerful alliances to dismantle the master's house. In line with this practice, it is essential that disabled and chronically ill people speak out in ecological and transfeminist battles to broaden the voice of those who defend the earth and the living. We know how important it is to free care work from feminisation through collectivisation, in order to affirm all the different forms of what we mean by and desire about health and reciprocity. Through this process we can make the spaces of militancy and struggles more inclusive for everyone. This is the challenge against a world that portrays us as passive and silent victims if not unproductive and unworthy lives.

Infection

> ***"Be careful when you go down there, it's dangerous, don't wear short pants"*** –This was the friendly advice of a woman representing the company in charge of mosquito control in the region of Skala, Greece, where malaria had taken hold during the years 2009–2015. At first, I thought her comment was referring to avoiding getting bitten by mosquitoes in the orange fields I was going to visit. Moments later, I realised she was referring to protecting myself from male farmworkers from Pakistan and Bangladesh. Reproducing racist and colonial imaginations of dark-skinned people as hypersexual, uncivilised, and savage, this pretension of woman-to-woman solidarity made my stomach turn.
>
> —Panagiota Kotsila, personal notes

Panagiota: "Farmers here used to gamble their cars and houses during the 'good times,' in the casinos. A lot of fortunes made and lost down there, all from the oranges", my father said as we were sipping coffee at the porch of the house where he grew up, at a village near Sparta, Greece. It was 2015 and he had not yet heard about the rise of malaria cases in this region since 2009, which remarkably coincided with the beginning of the economic downturn and what would be a long period of crisis-induced austerity. He was also unclear about why I was visiting the town of Skala, in the agricultural plain of the Evrotas River Delta, an hour and a half drive away—to research land change and agricultural practices, mosquito ecologies, or the living conditions of immigrant farmworkers? Here too, microbes, mosquitoes, oranges, land, immigrants, bodies, farmers, political economy, and public health "are all part of a single story" (Mansfield, 2011).

My family used to own orange trees in the region. Now we only have some olive trees, as they demand less water and attention. "Olives are sacred", my grandma used to say. She became a widow only a year after she was disinherited by her family for falling in love with my grandpa. She used to collect wild oregano and mountain tea as a source of income to raise her only son. Since my father moved to Athens, the management of the fields is delegated to a local family who in turn hires farmworkers to complete the harvest every winter, getting in return half of the resulting organic, cold-pressed olive oil from the local cooperative. I seem to come from a line of local producers in the region, of hard-working people working the land, of a single mother who struggled to survive.

About two weeks before, I was sitting across the desk of a public health worker in the Hellenic Center for Diseases Control and Prevention, back in Athens. She was rather confidently describing the success of the anti-malaria program in the Evrotas region, through door-to-door visits, body temperature monitoring, drug administration, distribution of bed nets and anti-mosquito plug-ins to the farmworkers living in the fields. At my mention of the deep discriminatory structures and obstacles that immigrant farmworkers face in Greece, she resorted to self-defensive claims: "We tell the farmers to protect their (sic) immigrants, to give them anti-mosquito appliances and take them to the doctor". I seem to come from a society where the dehumanisation of immigrant subjects is deeply ingrained.

A month later in early August, I was walking around the orange fields in Skala, looking for the shacks and half-finished constructions where immigrant farm workers lived. As a river delta, the Evrotas region is fertile and supports a vast cultivation of oranges and other agricultural goods. The seaside swamps, lagoons and sandy beaches, coupled with an expansive network of irrigation canals, make the region also a prime location for the *Anopheles* mosquito, a carrier of malaria. Since the late 1990s, male migrants mostly from Pakistan and Afghanistan work in these fields. Some 800–1,000 workers reside there permanently while others only come for the harvesting season. In 2010, the number of immigrants in the region peaked at 1,500, with most of them being irregular and risking deportation under an aggressive immigration policy. In 2011, there were 42 domestically acquired malaria cases documented—the highest reported since 1974.

The first years of rapid malaria spread in the region of Skala (2009–2011), the issue remained concealed in internal reports of local authorities and did not make local news (Kotsila & Kallis, 2019). For the (government-friendly) media, it only became a story worth telling in order to cast blame. Enhancing a neoliberal personification of health risks, vulnerable individuals were presented as the risk themselves, because they were mostly affected by it. During those years, in a context of severe cuts in public healthcare and other welfare sectors, immigrants, sex workers and people with drug addiction were bluntly blamed for the rise of infectious diseases in Athens (Kotsila & Kallis, 2019).

During the first years of spread, malaria was a non-issue for the national public health authorities, as it appeared to mostly affect immigrant groups in the region. But, in 2012, when the number of Greek citizens infected also started to rise, and the tourism sector was facing the risk of a red flag from the World Health Organization, it became a reality worth addressing. Immigrant bodies in Skala were no longer perceived solely as moving the national "productivity wheel" through their cheap labour, but also, at the same time, as potentially blocking it. Immigration was then treated as a disease risk factor, rather than as a site of vulnerability. The biopolitical caring state suddenly "saw" immigrant farmworkers and targeted them as subjects of public health interventions, including enforcing preventive medicalisation only for immigrants, often without their clear consent.

This operated within a broader and historical bio-/necro-politics of immigration (Foucault, 2003; Mbembe, 2020) through which irregular immigrants are discursively and practically treated as ambivalent subjects: a threat to the nation-state's "well-being", at the same time necessary as workers for many of its productive sectors. As such, both a risk and an asset, they are continually being managed and controlled. The malaria epidemic exacerbated this process of "an internal racism of permanent purification" (Foucault, 2003, p. 62) and a "constant exposure to conditions of death" (Mbembe, 2020), by tying already existing xenophobic attitudes and anxieties to concrete health aspects, as immigrants from malaria-endemic countries were blamed for introducing the disease to local environments.

Shirts and pants were drying in the sun on a rope between two trees. I was sitting outside a shack, strategically hidden among orange trees in the middle of agricultural land. These houses were as big as $10m^2$ and shared by 6–8 workers, and with no glass on the windows to keep mosquitoes out, still, four years after the malaria outbreak. Some of the men joined me around the table; others seemed to want nothing to do with a white girl holding a notebook and asking questions. They told me how they were not allowed to rent a decent home in the village, how they were owed salaries, and treated "like animals", how they were the last to be attended to in the health centres. "There is no way to protect from the mosquitoes, but more than getting sick, we are afraid of getting deported", said one of the youngest from the group. I doubted whether the older man, who kept a big stack of painkillers next to his bed, would agree—his body would not take a malaria infection, or maybe it had already been through it.

> A couple of years ago [2014], doctors came here to make tests and give us medicine. A couple nights later, the police came together with others. They beat us, broke our stuff, and arrested some. We couldn't trust the doctors again.

I suddenly realised that malaria was not the outcome of rational decisions about how to better protect from it (wear long clothes, use bed nets, avoid standing water), or of the politically defined spatial and temporal coexistence of orange trees, irrigation canals, mosquitoes and people's bodies. Similar to what Doshi (2017) has highlighted for urban contexts, here also it was the embodied, emotional, and visceral realities of state violence, social exclusion and fear that pushed farmworkers to live in the fields and avoid public health authorities, tests and medical visits, even if it meant that the epidemic from which they were also suffering would be harder to control. In Skala, just as in most contexts of intensive agricultural production of high-value (mono)crops, immigrant farm workers often lack legal recognition and live precariously from one season to the next. They are not considered part of the population socially or politically—they provide labour but have very limited (labour, health, civil) rights (Agamben, 1998). While targeted public health interventions might have helped fight malaria, they did not question the broader structures of racial capitalism and enduring colonial ideologies that had been shaping immigrants' livelihoods and health for years in Skala, and beyond.[2]

I had spent countless summers in rural Peloponnese, but home for me is the urban centre of Athens, where I lived and studied for most of my life. During my fieldwork, I was a local, but I also was not. Living abroad for seven years, I even felt disconnected from the reality of the deep economic crisis in Greece at the time. I was also a migrant, but one well shielded behind white skin, a European passport, and highly respected institutions. It was these felt "ambivalences, discomfort, tensions and instabilities of subjective positions" (Sultana, 2007, p. 377) that pushed me to tell this story about the margins of recognition, citizenship and belonging. It is there where I found the hands that pick, the bodies that bend, and the faces that—after days of listening and engaging with their stories—smiled back to me.

Nausea

> ***"So that's the process of the whole Zofran pump...I'm in the military so, I had to figure out how I'm going to carry this pump around..."*** A vlogger explained via vlog how she experienced Hyperemesis Gravidarum (HG) and how she connected her Zofran pump[3] to her pregnant

[2] Neely (2021) speaks of this in the context of health development programs in the global South. Here, we see one way in which these relationships are reproduced in the context of enduring colonial ideologies in Europe.

[3] A 'Zofran pump' refers informally to a subcutaneous (SQ) micropump device for infusing controlled doses of a medication (in this case Ondansetron, which is commonly sold under the brand name Zofran), which ensures faster uptake of the medicine than oral modes of delivery (especially when a person is unable to tolerate swallowing much of anything), while allowing a person greater mobility and independence than with larger, intravenous treatments.

belly and carried it with her in uniform at her workplace. I watched her vlog among many others on YouTube several times and read through the comments of hundreds of others struggling with HG. Both HG vlogs and Zofran were life-altering technologies carrying me through my pregnancy with HG. They exist through corporate infrastructures mediating everyday social relations, *and* they enabled me to leave my bed to continue teaching undergraduate students and to bring my pregnancy to full term. This uneasy reliance on serotonin 5-HT3 receptor antagonist drugs and social media voyeurism pushed me to empathize beyond my own embodied experiences and to confront my own ableist understanding of various technologies as an early career feminist political ecologist.

—Ingrid Nelson, personal notes

Ingrid: It is uncomfortable for me to admit that both the drug Zofran and watching hours of HG vlogs helped me to leave my bed, carry my pregnancy to term and keep my job. While data on HG is woefully rare and unevenly gathered, the Hyperemesis Education & Research (HER) Foundation[4] estimates a "fetal loss rate" of 34%, and more than half of those with HG experience job loss. I understand these drug and social media social technologies on which I relied with thankfulness, ambivalence and as troubling, augmentative and largely unseen in many discussions of pregnancy in my personal and academic settings.

I spent the fall season of 2017 bed-bound with hyperemesis gravidarum (HG) until my insurance provider finally approved the off-label use of Ondansetron (commonly known and sold as Zofran), to enable me to eat again, and to leave my bed to continue to teach undergraduate students. I had lost twenty pounds in my first trimester, was vomiting between ten and twenty times per day and nearly required hospitalisation due to dehydration. Hoping to avoid the occasional foetal heart complications that can occur with taking Ondansetron in the first trimester, my obstetricians initially dismissed my symptoms as me being too wimpy to handle common morning sickness. As a member of my faculty union in the context of the United States, where access to health care is not yet a right, I was privileged to access health care benefits in my union contract that paid a large portion—though not all—of the cost of my medical care. I tried anti-nausea wrist bands, an expensive new drug called Diclegis, and

[4] https://www.hyperemesis.org/about-hyperemesis-gravidarum/.

attempted to follow any advice I could find in general pregnancy information materials in books and online to no avail. I remember staring at the bottle of Zofran pills before gratefully taking my first dose, which also triggered intense memories of helping my mother keep track of all of her pills—including her Zofran—during her radiation and chemotherapy treatments for cervical cancer when I was a teenager. She fought hard to access this expensive serotonin 5-HT3 receptor antagonist.[5] It blocks the working of serotonin in the part of the brain stem responsible for the involuntary vomiting reflex. I remembered my parents arguing about how disgustingly expensive treating cancer can be.[6]

Before swallowing Ondansetron, I was too weak to do much of anything, including watching online videos or engaging with social media. I had to teach two days per week. My partner drove me to my first class, where I sat and drew diagrams and key terms on a "doccam" projected onto a screen; I couldn't stand and write on the board. I then made my way to my office where I slept on the floor until I had to go to my next class, an advanced FPE seminar in which students co-designed and co-led discussions centring our readings for the day. My partner would then pick me up and I would remain in bed—with the exception of bathroom visits or being driven to the obstetrician's office—until the last day of classes for the week. Ondansetron made it possible for me to move about in a campus landscape that I otherwise could no longer access due to weakness, dehydration, sudden weight loss, and nausea. Even with this drug, I still vomited most days, just less often. I wasn't able to eat a "full" meal until after giving birth. Once I had the drug, I also mustered enough energy to at least resume more active teaching. I also started to engage more with social media where I found vloggers who described their experiences with HG.

Most HG vloggers were careful to provide disclaimers that they were "not a doctor" and encouraged viewers to seek professional medical advice if they suspected they might have HG, even as most complained about their doctors not believing them for an unacceptable period of time. Many themes emerged in the vlogs, such as how to distinguish

[5] In 1998, her bottle of ondansetron cost nearly US$ 1,200.00. Generic versions only became available in 2007.

[6] I deeply appreciate S. Lochlann Jain's 2015 critical analysis of the cancer industrial complex, a book I have taught in an undergraduate seminar four times with profound impacts, as expressed in various student feedback.

between morning sickness and HG, which results in drastic weight loss and dehydration and does not respond to morning sickness support measures. Others demonstrated how to set up and attach Zofran pumps and others narrated their experience while hooked up to IV lines in hospital beds. Several vloggers addressed losing a pregnancy or resorting to abortion while others addressed finding support to care for toddlers while experiencing HG. Many spoke of financial hardship, job loss and challenges with partners or others in "support" networks who dismissed the severity of their symptoms. These vlogs broadened my understanding of HG across age, race, class, sexuality and other categories. Here, those whose complaints are especially disregarded by medical practitioners, such as women of colour and queer women, wait longer for diagnosis and to receive treatment, if they do. Those without childcare support and who cannot afford time away from work during pregnancy also struggle, centring class and caregiving contexts as critical. There were major differences across class, childcare and work contexts in terms of who experienced job loss or lost the ability to care for their children while pregnant. They also demonstrated the range of severity of symptoms, with some women being hospitalised for long periods, others relying on portable Zofran pumps and others managing with pills while still others experienced resistance to most anti-emetic drugs.

Media studies scholar, Samira Rajabi (2021, pp. 70–71), argues, "The Internet, particularly in the case of traumatic suffering, has been shown to take passive sufferers and enable them to make themselves into active media prosumers, engendering new points of significance and fostering the potential for change...whether the trauma was from experiencing, looking, or feeling empathy, the process of meaning making functioned in the same way", in the context of a woman's viral breast cancer photo story created by her husband. The HG vlogs strengthened my empathy for others with HG while providing me with context and thankfulness that my Zofran pills were working enough to keep me out of the hospital while not necessitating a pump or other additional support. This broader perspective also made various remarks by my obstetricians about how I shared something with select celebrities all the more infuriating. Unlike attempting to find experiential connections between myself and a famous person, viewing multiple HG vlogs highlighted illness as always situated in a social, place-based and political-economic context. The vlogs highlighted the essential roles of family and other caregivers, insurance companies and everyday mundane practices shaping experiences of HG.

Many vlogs "disappear" over time or are hard to locate long after they go online. At the same time, other digital presences persist beyond when a vlogger might want them to. HG is both temporary (for the duration of a pregnancy) but also lingers in the sense that since having my daughter I have refused to become pregnant again. Every mention of a second child and the joys that a sibling might bring to my daughter provoke a visceral reaction in my body—never again will I go through that experience, especially knowing that HG is quite likely and often increases in severity with every pregnancy. Some people develop resistance to the drugs. I just can't *imagine* myself enduring a pregnancy without the drugs that allowed me to eat instead of wasting away in the hospital, this time with a young child to care for. Some with HG decide to have more children, especially if they feel supported and if they know they can access the drugs they need. I would likely need to take unpaid medical leave and that seems unaffordable. Some of my Angolan partner's extended family see me as a failure for not having many more children. The accusations amount to, "See what you get for choosing a white wife!" when really in this case this is "what you get" with someone who gets HG. Or maybe I wouldn't get HG a second time around, but the odds are not good. There is a complex temporality to these kinds of calculations.

In the case of HG, temporality also complicates the building of virtual communities and notions of crip identity politics. Many who experience HG can "move on" after their experience because HG has an end (the end of pregnancy). Within 3 hours of giving birth, I ate my first full meal: a greasy grilled cheese sandwich. Although it gave me some temporary heartburn, the speed with which my body shifted from an aversion to most food (even with the drug) to no more nausea was striking. Three hours! Within the next week, I became ravenous as my body suddenly transitioned into needing to produce breast milk while also gaining back all the weight I had lost during my pregnancy (the opposite situation for many). While there are some who take on HG as a cause to support and to build community around supporting others. This can also get infused with challenging abortion rights politics. Some anti-abortion activists see HG as an instance where they can actively coach pregnant people away from abortion. The majority of vlogs I encountered avoided questions of abortion or framed it as something for the pregnant person and their doctor to work through. There is a plethora of community and activism in these digital spaces, but I wouldn't call it all good activism or feminist.

Regarding HG activism and temporality, for how long does "response-ability" apply? With whom are we in community, for how long and why? Are short-lived communities of care and activism also laudable? Feminist? What happens to an issue/disease/experience as something worth organising around, if it's not "chronic" (although it is likely to repeat when pregnant again), and thus, not easy or obvious to build an identity around?

Discussion

Porosity and the Symptoms of Relational Ecologies

How do these stories speak to the porous, inextricable relationships between bodies and places; bodies and territories?

Chronic pain is a symptom of the porous relationship between body and place, even though it is often silenced in the biomedical sciences. For Ilenia, the embodied and everyday experience of pain sparked the desire to learn from the composite history of struggles for social and environmental justice animated by Black, Latinx, Indigenous and proletarian women and LGBTQIA+ folks (Anzaldúa & Keating, 2002; Kafer, 2013; Lorde, 1980; Pulido, 2008). Their activism employs the politicisation of ill bodies (Iengo & Armiero, 2017) as a peculiar kind of situated knowledge reflecting on how the health conditions of a community have much more to do with the accumulation of institutionalised oppression and toxicity than with a personal fault or incapacity (Iengo, 2022). It took years of days filled with pain, sometimes normalised, or at best met by a sense of pity by a myriad of medical doctors, teachers, friends and lovers, to come to terms with the relevance of pain in understandings of intersecting environmental, social, gender oppressions. A cripistemology of pain contributes to affirming the experiences of pain and the situated, partial knowledge that stems from such embodiment (Patsavas, 2014). In situating the experience of chronic pain within the cultural politics of ableism, a cripistemology of pain criticises the reduction of chronically ill and disabled people's lives to body parts that hurt.

Ilenia's chronic inflammatory pain functioned as an instigator to politicise the porosity of embodiment. In the Land of Fires, it means engaging with the uneven infrastructure of chemical relations where dioxin bioaccumulates into human and more-than-human bodies as a result of hazardous waste burning (Iengo, 2022; Murphy, 2017a, 2017b). In this reflection,

Ilenia thinks with the proposition by Sara J. Grossman to be attentive to such relationships of pain and narrative that leak between biospheres and bodies (2019). FPE needs to make space for the cripping and queering work around the politics of knowledge reproduction on socio-environmental health, which go beyond the victim narrative or that of somewhere/someone deemed unnatural, sick, unproductive and in need of fixing. FPE has a long history of amplifying autonomous and grass-roots efforts in producing knowledges that can help us interrogate issues such as the feminised and sexualised nature of care work which does not make space or create possibilities for the sick woman. Ilenia proposes a cross-fertilisation of transfeminism, critical disability studies and environmental humanities in fostering perspectives that move beyond the normative assumptions often found in the biomedical sphere.

Together with comrades and friends, she recalls discussing: What happens when the body who needs care/to be attended to is the one naturally charged with the responsibility of care? A short circuit is the most fitting metaphor, because the entire relational, social, economic system on which the patriarchal society is based upon is the paid, underpaid, unpaid, naturalised and unrecognised labour of women and feminised subjectivities. If the disease keeps her in bed, unable to do much, who will take care of her, what kind of future can she imagine? These are the questions that keep her awake at night, when the pains are not so acute and there is room for thought. There is still a lot we need to do in our political communities to be able to think and practice response-ability towards the vulnerability, care and desires of our non-conforming bodies.

Through a different material experience, in the case of malaria in the Evrotas River Delta, the disease itself was the outcome of and the political ground on which a crisis of biopolitics was articulated and built (Kotsila & Kallis, 2021). This process illuminated how undesired, but otherwise "natural" ecologies, such as those of mosquito populations in wetland environments, become one of the outcomes of state failures in environmental management and result in disease. At the same time, these undesired disease vectors also become the place—the pore itself—of the porous relationships between places and bodies. The fact that those most exposed and affected by this relationship have been immigrant farmworkers makes visible the deeply political processes that underlie the decisions around who, under which conditions and with what consequences, will become entangled with such "undesired" natures. Here, it is not only how agricultural production will take place and whether it will be sustainable, organic, and local that matters; it is mainly a question of agency, ownership, rights and ultimately social and political power that defines the condition of agricultural workers and their everyday experience of enduring both undesired ecologies and status of undesired/denied citizenship.

Health inequalities are determined socially (Marmot, 2005). Malaria, specifically, demonstrates the need to examine disease under a more holistic and critical framework of socio-environmental evolution and change, as it depends on the intersecting ecologies and geographies of human, mosquito and water bodies. As Carter (2012, p. 2) has noted in Argentina: "malaria thrived in the region's hot, humid, subtropical environment but also flourished opportunistically in bodies worn down by alcoholism, malnutrition, overwork, and material deprivation".

Immigration itself, in this sense, can be considered a social determinant of health (Castañeda et al., 2015). The recent work of Teresa Mares (2019) on the dairy industry in Vermont, US, for example, also points to why occupational health risks and the non-access to health care impose tremendous challenges for immigrant workers in their efforts to care for themselves and their dependents. In Skala, too, as Panagiota here explained, malaria spread not only because immigrants did not have access to health care, but because they were working in highly mosquito-prone areas, inhabiting humid orange orchards with irrigation canals, in substandard accommodation and with no access to means of disease prevention. In turn, this was tightly linked to their status as non-citizens and their racial and ethnic background. As Carney (2014) notes, intersectional analysis has to include this experience of "illegality" as an axis of barriers to resource access and "the embodiment of subordinated status" experienced by undocumented migrants (citing Quesada et al., 2011, p. 351)".

It is thus important to think of these everyday embodied realities of experiencing, inhabiting and becoming in relationship to food production—and of any process of human–environment interaction and relation—in order to move away from romanticised visions of places or processes that might appear idyllic and pure in ecological, and as an extension, ethical terms. Although in biological terms, the interaction between bodies and environments are governed by the same biophysical "laws", the co-becoming of farm and farmer, of earth and earth-carer, of food and food producer, is always conditioned upon intersectional identities and positions of oppression or privilege. We need to speak more of that in our analysis of socio-environmental realities and future transformations.

As Ingrid explains, the aetiology of HG is still largely unknown, but it is broadly understood as "internal" in cause, rather than the result of exposure to a chemical such as dioxin or a plasmodium parasite...although HG occurs only while a person is pregnant (jokes and feminist writing about foetuses as kinds of parasites abound by the way), and then ceases within mere hours after the end of a pregnancy. Here, Ingrid sees fruitful scholarship from thinkers such as Sophie Lewis (2018, 2019) on gestational justice and the concept of "full surrogacy" as a way to reconfigure the connections, politics and responsibilities between people's bodies, social connections and places. The shift in her hormones also altered her connections with place through smell. When experiencing HG, her nausea was easily triggered by both commonly recognised foul odours and by mundane smells not normally considered illness-inducing. To this day, the seasonal change to autumn where she lives and works, in Vermont, brings scents of fermenting apples fallen from trees, a variety of fall flowers and seasonal foods, which triggers nausea long after having had HG. Autumn was the time in her pregnancy before her insurance approved Zofran, when she rapidly lost the most weight and was the least mobile. Her brain apparently still associates the scents of this season with the worst of HG for her.

Technologies and the Politics of Visibility

In the case of endometriosis and chronic pain, Ilenia describes her experience of health-related technologies, or the role of technology in experiencing and navigating health conditions, which reveals also how digital technologies can help visibilise and empower in a subversive, emancipatory way. In 2017, she started to search for an online crip community while her symptoms became unbearable, until she was stuck in the house or in bed for weeks or months. In an anglophone context, she found people who told their story, unrecognised by medical staff, others who provided everyday life hacks and advice for managing symptoms, and others dedicated to divulging the latest scientific advancements in simple and accessible language. Where the health professionals had failed in recognising what was wrong, there were so many people who according to the feminist practice "hermana yo sì te creo" ("sister I do believe you"), not only validated symptoms but were filled with anger at how the medical system treats chronically ill women, BIPOC and non-binary people, and at the same time offered empathy and support to those who took the first steps in the community.

Thanks to these online networks, archives of materials to study the complexity of the disease in depth are made accessible, to help awareness-building around the possible therapeutic options, against the oppressive medical myths to be debunked,[7] and the rights that we can affirm as patients. In Italy there is a flourishing of online dissemination and mutual help between people with endometriosis happening at the grassroots level. Info.endometriosi[8] and chroniqueers[9] are online support communities that divulge expert knowledge and create space for sharing experiences, hacks, and push the boundaries of what counts as knowledge about who is the endometriosis patient from an intersectional and transfeminist perspective.[10] Their work uses digital spaces strategically to build community, and overflows in spaces of grassroots organising with assemblies and self-training events.

[7] Endometriosis is systemic chronic inflammatory illness where endometrial-like tissue creates lesions, cysts and nodules, growing its own nerves and blood vessels causing inflammation, scar tissue, fibrosis and adhesions in extra-uterine context, especially in the pelvic and abdominal area, although it has been found on all organs of the human body. However, the vast majority of medical school training still casts endometriosis as a menstrual issue. For more on the sexist, racist, classist discourses in endometriosis science see: Capek (2000), Guidone (2020), and Iengo (2022).

[8] https://www.instagram.com/infoendometriosi/.

[9] http://chroniqueers.it.

[10] In the book, "The Makings of a Modern Epidemic: Endometriosis, Gender and Politics", author Kate Seear, traces how biomedical science has defined a prototype of person with endometriosis with complex consequences for those identified and those invisibilised. For decades, endometriosis was also called "the career woman's disease". Manifested in this definition is the construction of patriarchal, classist and racist medicine that obliterate BIPOC women and LGBTQIA+ folks. In the 1930s, physician Joe Vincent Meigs, concerned about declining birth rates in affluent, white communities in the United States, defined endometriosis as a consequence of prolonged, uninterrupted menstruation. Drawing parallels with primate females, whose menstruation is a rare occurrence between pregnancies, he proposed multiple and early pregnancies as a natural condition for female health. In addition, Meigs spent much energy in advising physicians to encourage their affluent and white patients to reproduce to overcome the population growth of the subordinate classes. To date, it is uncommon to hear such blunt arguments, although the ways in which "the endo patient" is classified and the solutions offered remain the same. An intersectional analysis allows us to unveil how the assumption that the disease affected white, upper-class women who choose a career over family came to be imposed, obliterating and keeping proletarian and racialized women and LGBTQIA+ completely out of the statistics. It takes an average of 7.5 years to receive a diagnosis. For black, racialized and LGBTQIA+ people this time is multiplied out of all proportion, and they face greater difficulties in accessing already risky healthcare, while experiencing more medical violence in "women's" wards.

Disability justice and crip activism mediated and flourishing via digital technologies has offered us tools to counter-narrate our own bodily experiences, against the pathologisation of the body and the patronising practices of the medical establishment, and to navigate anti-ableist everyday life practices as disabled and chronically ill folks. It is essential to talk about the transformative power of our crip communities without indulging in inspirational narratives of the "super crip" focusing on overcoming illness and achieving well-being *in spite of* the conditions (Clare, 2009; Hamilton, 2014; Schalk, 2016). Nobody can determine what quality of life means to us but us! The mutual aid networks between spoonies[11] have been fundamental for education towards awareness and improvement of symptoms, while reinventing our material and discursive worlds (Hamraie & Fritsch, 2019).

In the activism of chronically ill and disabled people, the issue of data is fundamental: where there is no information, it must be collected through self-organised inquiries and analysed from an intersectional perspective to fill those gaps in the representation of class, race, gender, nationality, sexuality and of course ability, to raise awareness on the lurking oppressions and injustices. Some have called the practice of producing this transformative, creative, knowledge that is specific to the experiences and lives of chronically ill and disabled people, as "epistemic activism" (Hamraie, 2017, p. 132) in opposition to the dominant forms of knowing and acting around disability. Expanding the narrative of contested illnesses allows us to explore together and engage in processes that help recover our dignity and to overcome mainstream discourses that portray people with disabilities and chronic illness as expendable. This work, which is often done via digital technologies, exposes how ability, gender, sexuality, race, class, nationality influence our experiences of biomedical, economic and environmental violence, and from there to think about the practices, technologies and emancipatory imaginaries that we can put in place. These experiences do not allow a linear and universal prescriptive definition of

[11] Spoonie is a concept that refers to the "theory of spoons" used in the world of chronically ill and disabled people as a metaphor and visual representation of the daily amount of physical and mental energies that are indispensable for carrying out various activities. The theory, proposed by Christine Miserandino who lives with lupus, allows us to simply visualize how each disabled or chronically ill person has a limited number of spoons per day to deal with various tasks including showering, cooking, leaving the house, participating in a demonstration, have sex and work.

our political paths, which on the contrary must take into account the complexity and multiplicity of experiences and desires.

In the case of immigrant farmworkers, as Panagiota explained, taking care of oneself is often synonymous to avoiding deportation, to be allowed to work and to claim a place in local society. This is no easy task, and it often means balancing between being visible to actors who could help provide this access to basic survival (employers, human rights organisations, intermediaries, fellow workers), and staying invisible to others who would threaten it. Health and well-being were inextricably linked to this balancing act. Malaria, and the technologies used for its documentation and control (Google maps for locating the sheds where immigrants lived in the fields, medical diagnostic and preventive medicine technologies that only were administered to immigrants), were in many ways understood and experienced as a pretext for further biopolitical control and violence against these communities. Indeed, during the malaria spread, there were more frequent invasions of police forces into immigrant homes: Fridays after work, and before being paid for the week, sending many to detention centres, from where some would be deported. During public health visits, workers did not want to admit they were sick, because that would often mean being denied work and housing, but also could not deny examination and treatment because they were afraid of the repercussions. Outcomes of infection were thus defined both by how public health and medical technologies were mobilised and directed specifically to immigrant farmworkers, and as a consequence, by the meanings infused in malaria itself: both by those who portrayed it as a "backwards" disease carried by "backwards" subjects, and by those who suffer it in silence because they are embedded in contexts of exploitation and exclusion (Evered & Evered, 2012; Kotsila & Kallis, 2019).

As the other cases in this chapter also reveal, institutional oppression and control exercised through a biopolitics of health and related technologies, might go unnoticed by those with enough privilege to escape it. Already in the 1970s, the Italian physician and founder of Medicina Democratica[12] Maccacaro stated: "medicine is entrusted with the task of resolving, in scientific rationality, this contradiction of the capitalist mode

[12] Since the early 1970s Medicina Democratica has been a social movement that fights for the right to health in Italy. It was born in the working-class struggles and the fight against toxicity in the workplace, aiming to change the role of science and especially medicine towards creating knowledge that could be horizontal, collective and participatory.

of production, which on the one hand consumes and extinguishes labor power but on the other needs it to continue to feed itself" (1979, p. 140). While recognising the need for intersectional approaches to institutional exclusions from social welfare systems (Bowleg, 2020), we aim to move towards amplifying the knowledges and transformative justice praxes of those who are captured and oppressed by the state's medical governmental gaze (Khan et al., 2022). As Petchesky (2016) argues, "there is nothing disengaged, neutral or 'objective' in the knowledge or technics that biopolitics employs" (p. 171). Intersectional marginalised communities, from the non-citizen racialised men working in agricultural fields, to the working-class woman in the periphery of a metropolis, to the trans indigenous person with disability, experience in their flesh and bones that provisions of health often come with some sort of state and medical violence in its racist, patriarchal, and classist declinations.

Similarly, so, albeit entangled in a different kind of ambivalence with medical and online technologies, the two "technologies" that opened up access to place again for Ingrid were: one (the drug) by mitigating nausea and the other (the vlogs) by situating her experience among other pregnant people in different contexts, places, work obligations, etc. The vlogs in particular made her understand both the inaccessibility of certain aspects of her university campus and also how privileged she was with regard to the flexibility of work and the protections her contract offered as a member of a union. She has separate reflections on her experiences as a bisexual person in terms of the heteronormativity of the obstetrician's office, campus spaces, and even the vlogs, but she "expected" these kinds of silences and erasures (and funny moments) of the pregnancy. However, she did *not* expect having HG or how important a fraught drug and watching other pregnant people navigate the condition, explain how to use Zofran pumps and function at work with HG would be to even staying pregnant, keeping her job and avoiding deep depression. She is grateful for these technologies but also cognisant of their origins and broader troubling connections. Ingrid has lingering questions about how FPE should understand these technologies (see Nelson et al., 2022).

Conclusions

Through the lenses of the three embodied experiences of chronic pain, infection and nausea, we explored the themes of porosity and technology as they offer us ways to insist on the right to be and the politics of health in FPE. Building on feminist interventions in environmental humanities, we engaged with the porous relationship between bodies and places, their mutual as much as disabling exchanges and the onto-epistemological understandings that arise from this acknowledgement. The relationship between embodiment and medical/digital technologies challenge understandings of health as an individual characteristic and binary opposite of illness, moving towards narratives that account for health as the complex and ever-changing process deeply affected by access to mutual aid, care networks, medical and digital technologies (proprietary, reclaimed, hacked, never innocent), knowledge over our conditions and the processes of self-determination to repair harm and dismantle multiple systemic oppressions ingrained into our societies. The ambivalent nature of different kinds of technologies employed in and around health has been explored in their capacity to visibilise or invisibilise experiences.

Our chapter centred what is often at the margins of FPE discussions, concerning the kinds of porous relations between bodies and places that continuously permeate and inform politics, from the personal to the broader social level and the ambivalent meanings and usages of digital and medical technologies mediating our understanding and experiences of health and embodiment.

Funding: This chapter was funded by the Wellbeing Ecology Gender and cOmmunities Innovation Training Network (WEGO-ITN) funded by the European Union's Horizon 2020 research and innovation programme under the Marie Sklodowska-Curie grant agreement No. 764908-WEGO 2018-2021.

References

Agamben, G. (1998). *Homo sacer: Sovereign power and bare life*. Stanford University Press.

Anzaldúa, G., & Keating, A. L. (2002). *This bridge we call home: Radical visions for transformation*. Psychology Press.

Barca, S. (2014). Telling the right story: Environmental violence and liberation narratives. *Environment and History, 20*(4), 535–546.

Bailey, M., & Peoples, W. (2017). Towards a black feminist health science studies. *Catalyst: Feminism, Theory, Technoscience, 3*(2), 1–27.

Bowleg, L. (2020). Evolving intersectionality within public health: From analysis to action. *American Journal of Public Health, 111*(1), 88–90. https://doi.org/10.2105/AJPH.2020.306031

Capek, S. (2000). Reframing endometriosis: From 'career woman's disease' to environment/body connections. In S. Kroll-Smith, P. Brown, & V. Jan Gunter (Eds.), *Illness and the environment: A reader in contested medicine* (pp. 345–363). New York University Press.

Carney, M. A. (2014). The biopolitics of 'food insecurity: Towards a critical political ecology of the body in studies of women's transnational migration. *Journal of Political Ecology, 21*, 1–18.

Carter, E. D. (2012). *Enemy in the blood: Malaria, environment, and development in Argentina*. University of Alabama Press.

Castañeda, H., Holmes, S. M., Madrigal, D. S., DeTrinidad Young, M. E., Beyeler, N., & Quesada, J. (2015). Immigration as a social determinant of health. *Annual Review of Public Health, 36*, 375–392.

Clare, E. (2009). *Exile and pride: Disability, queerness, and liberation*. South End Press.

Da Costa, L. B., Icaza, R., & Ocampo Talero, A. M. (2015). Knowledge about, knowledge with: Dilemmas of researching lives, nature and gender otherwise. In W. Harcourt & I. Nelson (Eds.), *Practicing feminist political ecologies: Moving beyond the "Green Economy"* (pp. 260–285). Zed Books.

Doshi, S. (2017). Embodied urban political ecology: Five propositions. *Area, 49*(1), 125–128.

Evered, K. T., & Evered, E. Ö. (2012). State, peasant, mosquito: The biopolitics of public health education and malaria in early republican Turkey. *Political Geography, 31*(5), 311–323.

Foucault, M. (2003). *"Society must be defended": Lectures at the Collège de France, 1975–1976* (Vol. 1). Macmillan.

Grossman, S. J. (2019). Disabilities. *Environmental Humanities, 11*(1), 242–246.

Guidone, H. C. (2020). The womb wanders not: Enhancing endometriosis education in a culture of menstrual misinformation. In C. Bobel, I. T. Winkler, B. Fahs, K. A. Hasson, E. A. Kissling, & T. A. Roberts (Eds.), *Palgrave handbook of critical menstruation studies* (pp. 269–86). Palgrave Macmillan.

Guthman, J. (2011). Bodies and accumulation: Revisiting labour in the 'production of nature'. *New Political Economy, 16*(2), 233–238. https://doi.org/10.1080/13563467.2011.542801

Guthman, J., & Mansfield, B. (2013). The implications of environmental epigenetics: A new direction for geographic inquiry on health, space, and nature-society relations. *Progress in Human Geography, 37*(4), 486–504.

Hamilton, A. (2014). Working out some issues: The latest in 'inspirational' ableism. *Bitch, 62*, 5.

Hamraie, A. (2012). Universal Design research as a new materialist practice. *Disability Studies Quarterly, 32*(4). https://doi.org/10.18061/dsq.v32i4.3246

Hamraie, A. (2017). *Building access: Universal Design and the politics of disability*. University of Minnesota Press.

Hamraie, A., & Fritsch, K. (2019). Crip technoscience manifesto. *Catalyst: Feminism, Theory, Technoscience, 5*(1), 1–34.

Haraway, D. (2016). *Staying with the trouble: Making Kin in the Chthulucene*. Duke University Press.

Hedva, J. (2018, 17 January). *Letter to a young doctor*. Triple Canopy n. 24. Retrieved March 10, 2022 from https://www.canopycanopycanopy.com/contents/letter-to-a-young-doctor/#title-page

Iengo, I. (2022). Endometriosis and environmental violence: An embodied, situated ecopolitics from the land of fires. *Environmental Humanities, 14*(2), 341–360.

Iengo, I., & Armiero, M. (2017). The politicization of Ill bodies in Campania, Italy. *Journal of Political Ecology, 24*(1), 44–58.

Jackson, P., & Neely, A. H. (2015). Triangulating health: Toward a practice of a political ecology of health. *Progress in Human Geography, 39*(1), 47–64.

Jain, S. L. (2013). *Malignant: How cancer becomes us*. University of California Press.

Jaquette Ray, S., & Sibara, J. (Eds.). (2017). *Disability studies and the environmental humanities*. University of Nebraska Press.

Jones, C. E. (2016). The pain of endo existence: Toward a feminist disability studies reading of endometriosis. *Hypatia, 31*, 554–571.

Kafer, A. (2003). Compulsory bodies: Reflections on heterosexuality and able-bodiedness. *Journal of Women's History, 15*(3), 77–89.

Kafer, A. (2013). *Feminist, queer, crip*. Indiana University Press.

Khan, Z. H., Iwai, Y., & DasGupta, S. (2022). Abolitionist reimaginings of health. *AMA Journal of Ethics, 24*(3), E239-246. https://doi.org/10.1001/amajethics.2022.239

Kim, J. B. (2017). Cripping East Los Angeles: Enabling environmental justice. in Helena María Viramontes's "Their dogs came with them". In S. J. Ray & J. Sibara (Eds.), *Disability studies and the environmental humanities: Toward an eco-crip theory* (pp. 359–380). University of Nebraska Press.

Kotsila, P., & Kallis, G. (2019). Biopolitics of public health and immigration in times of crisis: The malaria epidemic in Greece (2009–2014). *Geoforum, 106*, 223–233. https://doi.org/10.1016/j.geoforum.2019.08.019

Kotsila, P., & Kallis, G. (2021). Politics of austerity and 'otherness' as public health determinants: The malaria epidemic in Greece during crisis (2009–2014). In R. Calvário, M. Kaika, & G. Velegrakis (Eds.), *The political ecology of austerity: Crisis, social movements and the environment* (pp. 92–111). Routledge.

Lewis, S. (2018). Cyborg uterine geography: Complicating 'care' and social reproduction. *Dialogues in Human Geography, 8*(3), 300–316.

Lewis, S. (2019). *Full surrogacy now: Feminism against family*. Verso.

Lorde, A. (1980). *The cancer journals*. Aunt Lute Books.

Maccacaro, G. A. (1979). L'uso di classe della medicina. In *Per una medicina da rinnovare, Scritti 1966–1976* (pp. 406–434). Feltrinelli.

Mansfield, B. (2011). Is fish healthy food or poison? Farmed fish and the material production of un/healthy nature. *Antipode, 43*(2), 413–434.

Mares, T. M. (2019). *Life on the other border: Farmworkers and food justice in Vermont*. University of California Press.

Marmot, M. (2005). Social determinants of health inequalities. *The Lancet, 365*(9464), 1099–1104.

Mbembe, A. (2020). *Necropolitics*. Duke University Press.

McRuer, R. (2002). Compulsory able-bodiedness and queer/disabled existence. In S. L. Snyder, B. J. Brueggemann, & R. Garland-Thomson (Eds.), *Disability studies: Enabling the humanities* (pp. 88–99). Modern Language Association.

Mingus, M. (2010, January 22). Interdependency (excerpts from several talks). *Leaving Evidence*. https://leavingevidence.wordpress.com/2010/01/22/interdependency-exerpts-from-several-talks/

Murphy, M. (2017a). What Can't a Body Do? *Catalyst: Feminism, theory, technoscience, 3*(1), 1–15.

Murphy, M. (2017b). Alterlife and decolonial chemical relations. *Cultural Anthropology, 32*(4), 494–503. https://doi.org/10.14506/ca32.4.02

Neely, A. H. (2021). *Reimagining social medicine from the south*. Duke University Press.

Nelson, I. L., Hawkins, R., & Govia, L. (2022). Feminist Digital Natures. Environment and Planning E: Nature and Space. https://doi.org/10.1177/25148486221123136

Nixon, R. (2011). *Slow violence and the environmentalism of the poor*. Harvard University Press.

Patsavas, A. (2014). Recovering a cripistemology of pain: Leaky bodies, connective tissue, and feeling discourse. *Journal of Literary and Cultural Disability Studies, 8*(2), 203–218.

Pellow, D. N. (2018). *What is critical environmental justice?* Polity Press.

Petchesky, R. P. (2016). Biopolitics at the crossroads of sexuality and disaster: The case of Haiti. In T. Schrecker (Ed.), *The Ashgate research companion to the globalization of health* (pp. 191–212). Routledge.

Piepzna-Samarasinha, L. L. (2018). *Care work: Dreaming disability justice*. Arsenal Pulp.

Price, M. (2015). The bodymind problem and the possibilities of pain. *Hypatia, 30*(1), 268–284.

Pulido, L. (2008). FAQs: Frequently (un)asked questions about being a scholar activist. In C. R. Hale (Ed.), *Engaging contradictions: Theory, politics, and methods of activist scholarship* (pp. 341–366). University of California Press.

Rajabi, S. (2021). *All my friends live in my computer: Trauma, tactical media, and meaning*. Rutgers University Press.

Sadler, J. (2017). War contaminants and environmental justice. In S. Jaquette Ray & J. Sibara (Eds.), *Disability studies and the environmental humanities. Toward an eco-crip theory*. University of Nebraska Press.

Schalk, S. (2016). Reevaluating the supercrip. *Journal of Literary and Cultural Disability Studies, 10*(1), 71–86.

Seear, K. (2014). *The makings of a modern epidemic: Endometriosis, gender, and politics*. Ashgate.

Sultana, F. (2007). Reflexivity, positionality and participatory ethics: Negotiating fieldwork dilemmas in international research. *ACME: An International Journal for Critical Geographies, 6*(3), 374–385.

CHAPTER 5

Ageing and Feminist Political Ecology

Constance Dupuis and Nanako Nakamura

Introduction: Ageing and FPE

What can making *miso* tell us about wellbeing in later life? How can caring for the environment coincide with caring for/with older generations? How can we take the multiplicity of ageing experiences seriously? We explore these questions in this chapter through a dialogue on our fieldwork experiences in Japan and Uruguay, during which we both investigated the intersection of ageing and environment. We bring to this discussion the insights of feminist political ecology (FPE) on intersectionality, socionatural relations and everyday practices in order to tease out the complexities of ageing experiences and to help better understand later life.

C. Dupuis
Erasmus University Rotterdam, Rotterdam, The Netherlands
e-mail: dupuis@iss.nl

N. Nakamura (✉)
Wageningen University, Wageningen, The Netherlands
e-mail: nanako.nakamura@wur.nl

W. Harcourt et al. (eds.), *Contours of Feminist Political Ecology*, Gender, Development and Social Change,
https://doi.org/10.1007/978-3-031-20928-4_5

Our starting point is that old age is not a state at which we one day arrive, in the way that retirement age is reached. You do not one day wake up old.[1] Rather, it is an inherently fluid and ambiguous process that takes place across the life course. Getting old is not easily mapped onto chronological age; as Rosario Aguirre Cuns and Sol Scavino Solari argue, "being old appears as a homogenised event due to the feature of having lived many years, to the decrease in ability (biophysical) and to the proximity to death. This centrality of chronological age in the representation of old age is naturalised in common sense and obscures the inequalities, differences and specificities of the social production of these groups" (Aguirre Cuns & Scavino Solari, 2018, p. 22). Instead, we examine the nuances in different embodiments of ageing as a way to look at counter-hegemonic responses to ideas of ageing that take place throughout one's life.

Ageing and aged experiences are relationally materialised in different places, spaces, and environments in everyday practices (Katz, 2018). How, where, with whom and with what we age is shaped by race, gender, class, ability, sexuality, place and other important dimensions of difference. This is where, as authors, we are conscious of the difficulty to speak across contexts as socially and culturally diverse as Uruguay and Japan. We are aware that the meanings and experiences of ageing are shifting and diverse. It may be more accurate to speak of ageing in the plural, as highlighted in the title of Aguirre Cuns and Scavino Solari's "*Vejeces de las mujeres*", meaning "The old ages of women" in Spanish.

To add to the complexity, ageing experiences are not only contextual but also influenced by normative ideas about ageing that have become dominant globally. Discourses of successful ageing, active ageing and healthy ageing circulate at policy and public health levels (World Health Organisation, 2002, 2020). These discourses reinforce ideas of homogeneity in later life, masking the intersection of social differences and inequalities (Katz & Calasanti, 2015; Sandberg, 2015).

Japan and Uruguay, while contextually diverse, are also subject to the dominant norms of ageing. One commonality is that these normative discourses set up clear binaries between what is desirable, good or healthy

[1] Though we are aware that the adjective 'old' often carries a negative or derogatory connotation, we are inspired by currents within social gerontology and critical ageing studies that are working to reclaim the term 'old' as a political project, underscoring the ways in which euphemisms can unintentionally reinforce ageism (Calasanti et al., 2006; Calasanti, 2008).

ageing and what is not. In neoliberal contexts, with the emphasis on the individual, ideas of successful ageing can be used to justify the retrenchment of social and health services in the name of older people's autonomy (Dillaway & Byrnes, 2009). The focus on individuals leads not only to homogenising of ageing experiences, but responsibility for "success" or "failure" in ageing is placed on individual older people. A focus on individualism leaves structural and contextual factors to the side, reinforcing the reliance on autonomy and the binaries of abled and disabled (Rice et al., 2017; Sandberg & Marshall, 2017). In our dialogue, we aim to blur these binaries and highlight the co-construction of those policies and enacted powers that drive ageing people's agency in everyday life.

In order to go beyond ageing norms while remaining aware of context and agency, we look at two examples of collectives which focus on old age. The first is the *Suisha* in Japan, a rural women's traditional food business, and the second is Uruguay's national network for older people, REDAM (Red Nacional de Personas Mayores). We look at ageing in these specific contexts in order to illustrate the dynamics of social, political, and ecological relations. Our empirical studies look at situated ageing experiences to help diversify the understandings of ageing and later life and to broaden ageing and environment debates as part of the contours of FPE sketched in this book.

To our studies on ageing, we bring FPE understandings of everyday practices and ethics of care. We trace the complexity of ageing people's relations, including social and cultural constructions of ageing and intersectional thinking. We question existing power discourses around ageing by attending to different ways of encountering and caring for others in everyday practices. Applying FPE concepts such as socionature (Nightingale, 2017), which speaks to the mutually constitutive quality of the social and the natural, we look at the role of the environment in experiences of ageing. We take up the FPE approach to understanding the environment as more than the experience of nature in the rural and urban contexts, instead understanding the environment as a relationally constructed dynamic process where ageing and aged experiences shape everyday life as ageing bodies intersect with machines, fungus, politics and climate in specific processes and contexts. Our focus is on the points of convergence where everyday life is maintained, continued, and repaired—that is, everyday caring practices (Tronto, 1993).

Following this explanation of ageing and old age, we briefly explain our different theoretical interests before presenting our two cases. Nanako shares stories about *Suisha* in Japan and Constance about REDAM in Uruguay. We then discuss FPE, ageing and generational approaches before proposing ideas for an FPE research agenda on ageing.

Theoretical Approaches

Inspired by FPE's intersectional thinking, our case studies look at different struggles that older people engage in and how their responses go beyond the hegemonic framing of ageing. We bring our two cases into dialogue as we explore what we can learn across these different contexts. Although we both draw on experiences from our PhD research, inspired by and in dialogue with FPE thinking, our PhD projects have different theoretical moorings, which we now briefly introduce.

Nanako looks at challenges in a rural Japan women's local business as a part of their everyday practices, whereas Constance examines in Uruguay the struggles for ageing concerns to be taken seriously within sustainability discourses. Nanako applies post-capitalist thinking (Gibson-Graham et al., 2013) to the Japanese case to see interdependencies of humans and non-humans (Sato & Alarcón, 2019; Nelson, 2017; Nightingale, 2013) ageing together in everyday practices as a result of hybrid choices of profit and non-profit making activities. In the context of Uruguay, Constance draws on rupturist gerontology, a stream of ageing thinking that speaks to the need to break with hegemonic understandings of ageing in order to deeply see and understand the everyday experiences and agency of older people (Piña Morán & Gómez Urrutia, 2019). These theoretical understandings allow Constance to explore the different temporalities (Rhee 2020; Vasquez, 2020) at play within the Uruguayan case, as well as connect specific ageing and sustainability struggles to broader conversations about socionatures (Nightingale, 2018; Singh, 2017). We use these differently theorised positions to put older people into the conversations on situated, everyday experiences circulating in FPE research and practice.

Stories of Ageing

In the two narratives that follow, we each engage with the other's research. The dialogue we present here is based on a series of semi-structured interviews and the exchanges in a public presentation of our work in the Feminist Political Ecology Dialogues on Rethinking Age, Generation and Population.

Nanako: Japan, Kunma Suisha no Sato *(*Suisha*)*

The village of Water Wheel (*Kunma Suisha no Sato*) is a local women's business group in the semi-mountainous Kuma region in Shizuoka Prefecture, Japan. For more than thirty years, *Suisha* has been producing hand-made food products and selling them at a roadside shop and restaurant that they run, where local people can also sell agricultural products. The original entities of *Suisha*'s food business were local women's collectives, supported by diverse management bodies with different objectives, that commonly aimed at improving different ways of food preparation, farming, and other everyday practices through collective learning. Like many other rural women's local businesses formed in the 1980s in Japan, *Suisha* was supported by the local and national governments in response to rising voices for gender equality and the urgent need to react to the challenging situations in ageing and depopulating rural communities. Since its establishment in 1987, *Suisha* has been contributing to community development and revitalisation, for which the state government granted it an award in 1989.

One of the guiding questions for my research is what kind of interdependent relationships are found in the attempts made towards community development by local businesses. Growing up in the capital city of my home region, I am intrigued by the everyday practices in rural communities which may be overlooked by mainstream political discourses on rural revitalisation, practices such as artesian soba noodle and *miso* making. Because of *Suisha*'s business, many people from urban areas have visited this rural community, invigorating the local economy and enhancing livelihoods. Despite *Suisha*'s efforts for rural revitalisation, the situation confronted by the community is still challenging as ageing and depopulation continue. *Suisha*'s efforts and practices stirred up my interests: how do *Suisha* members embrace the situation with lingering challenges, while maintaining their businesses? Who are they, emerging from the fragmented pieces of everyday practices? In order to explore these questions, I undertook ethnographic fieldwork for two three-month periods in late 2019 and in 2021–2022 to explore the interdependencies built through different post-capitalist practices embodied by ageing rural women.

Suisha's central product is *miso*, fermented soybeans. Although mass-produced *miso* is available at supermarkets, *Suisha miso* is hand-made as it would have been decades ago, by each household for self-consumption. *Miso* production begins with making fermented malted rice and moves to cooking soybeans and mixing them with the malted rice. These processes usually take six days for one batch, with multiple batches in process at the same time. The fresh *miso* then goes through the fermentation process for about a year or two. The most care-intensive process is at the beginning, when steamed rice and a fungus (*Aspergillus oryzae*) are mixed and made into *kôji*, the fermented malted rice. In the process, temperature control is crucial in managing the speed of fermentation. In *Suisha*'s *kôji* manufacturing facility, the temperature is set at 35–40 °C depending on the fermentation process, and inside and outside temperatures. If successful, hyphal fungal networks form, developing root- or branch-like filamentous fungal mycelium. It is possible to discern nurtured *kôji* from its appearance, frequently described as "*kôji* blossoms are blooming".

However, producing blooming *kôji* at *Suisha* is not an easy process. Their old-fashioned and aged equipment does not function properly to warm up the fermentation room, so *Suisha* members cannot only rely on the machinery temperature control. To supplement this defect, *Suisha* members check constantly and see if *kôji* malt grows well. If not, they discuss whether they should change warm water for humidifying the fermentation room, or adjust the temperature settings, in the absence of reliable working technology. These decisions for fine adjustments are made collectively, based on their experiences. The experienced sensory notion was referred to as "*Kan (感/勘)* puter" jokingly by the current leader of *Suisha*, not a *com*puter but *kan*, which means both feeling and intuition in Japanese. She implied that their affective care is much more useful for the ageing process of *kôji* fermentation than inputs from the machine. It also signifies that the material relationship between the *Suisha* members with *kôji* is tied to other relationships with the aged apparatus. These relationships show that there is not only one way of caring. Instead, caring relationships emerge from contingency in which *kôji*, the machine, and *Suisha* members are maintaining everyday practices in fluid and ambiguous ageing processes. Experiencing these ageing processes becomes part of socionatural relations. Socionatures experienced through ageing bodies sharpen the sensibility of caring for others, making up for the shortcomings of supposedly predictable, rational measures usually provided by mechanised processes. The *Suisha* leader also illustrated her own ageing in relation to the materiality of the aged facility; "We, humans, get old. No wonder machines get old too". Different ageing bodies of humans and non-humans navigate and negotiate how to make decisions about care for *kôji*, showing that the processes of ageing influence how we maintain, continue, and repair in everyday life (Tronto, 1993). Togetherness in the different experiences of ageing is relationally embodied through interactions with other humans and non-humans (other members of *Suisha*, the *kôji*, the manufacturing equipment) that are ageing and surviving together.

Kôji is a hand-made product for *Suisha miso. Suisha* members pinch steamed rice using their thumb and ring finger to check the appropriate softness. Once it is ready, the fungus powder is added to the rice. *Suisha* members waste no time and start massaging the rice and the fungus powder together to make scratches on the surface of the rice grains and let the mycelium stretch around the network. The steamed rice is extremely hot at the beginning: when I joined in the activity, I needed to use a wooden rice scoop, whereas the skilled *Suisha* members quickly switched to using their hands to massage the rice. Meanwhile, the rice and *kôji* mixture cooled down to human body temperature. The relationship between *Suisha* members and *kôji* is an intimate socionature: "Your hands are getting smooth, aren't they? That's our beauty secret", a *Suisha* member told me while I had trouble scrambling the rice and *kôji* mixture. Due to several natural components of *kôji*, such as *Glucocerebroside* and amino acid, the cosmetic industry has researched its effects on skin regeneration or healing to harness this "beauty secret". For *Suisha* members, taking care of *kôji* is not only a process of producing food products, but also human and non-human interactions through affectionate touching, intuiting how *kôji* are blooming because of their care. In turn, the *kôji* care for aged human bodies. The ageing experiences of *Suisha* members are materialised in everyday practices based on ethics, through affective interactions with other humans and with non-humans (Puig de la Bellacasa, 2017) (Fig. 5.1).

Rural revitalization is an objective of *Suisha*'s umbrella organisation, a non-profit organisation (NPO) *Yumemirai-Kunma*. The NPO showcases their economic activities through running a roadside shop and restaurant, selling their products to stimulate the local economy. The NPO was established in 2000 to take over the leading role of the communal committee, the main group spearheading the communal revitalization project. Hoping to contribute to rural revitalization in ageing and depopulating communities, *Suisha* members transformed their business model, taking over management and supervision duties. Instead of choosing a capitalist enterprise, they adopted the form of NPO to be able to use the

Fig. 5.1 *Kôji* making—massaging rice and *kôji* malt

profit obtained through their food business for non-profit activities and the community. Unlike mass production companies, *Suisha*'s products, such as the *miso* based on hand-made *kôji*, gain added value by using traditional production methods that require more time and effort. Challenges and intimate human and non-human relationships that appear in the production processes rarely come to the fore in their rural revitalization story, even though vital responses made through encounters with others manifest behind the scenes. These activities do not directly lead to profit making, but produce other important values by maintaining traditional food practices and caring for ageing communities, which enrich their business profile.

The hybrid approach of profit- and non-profit making exemplifies attempts to enhance interdependent relations with others. As an NPO, *Suisha*'s food business engages with various other community members by organising seasonal festivals and touristic events. Most of the events showcase the nature-rich environment around the roadside station, attracting urban visitors. By doing so, NPO members expected to attract more people, potential customers of *Suisha*, to visit the communities and the station. *Suisha's* business as an NPO attempts to contribute to local economies and reinvigorate the languishing local atmosphere by encouraging increased communication among/between the locals and visitors.

Despite the attempts made by *Suisha* and the NPO towards economic revitalization through high-value food business, the situation remains challenging in ageing and depopulating rural communities. Still, what drives *Suisha* is rural women's ethics in attending to the ageing rural communities. Encountering others through business practices bridges different ways and bodies of ageing. The ethics speak to what they need to do; that is, the ethics of care emanate from the situated ageing rural women's perspective.

Ageing also evokes uneasy feelings about what the communities will look like in the future, a shared notion among the local people. This uneasiness motivates the entrepreneurship of the rural women, who aim to make economic relationships with others visiting from outside of the communities. The prolonged languishing local condition also drives women in *Suisha* to attend to the relationships *within* the communities. One member noted that they cannot carry on their community-based business without their communities. "I want to live in Kuma forever, and I think we want to keep living here even if one of us [husband or herself] is left alone. In the end, however, there might be no way to live here. But I want to set up a condition that we can live continuously". She joined the NPO as a member of staff in 2001 after she retired from her career as a kindergarten teacher. She is currently in charge of the lunchbox service for single elderly people and food trucks, one of NPO's non-profit-making activities. "I think newly joining members need to know the origin of *Suisha*", meaning how members from different women's groups orchestrated different initiatives together with the local governments, men and other women members in the communities, and visitors, with the shared commitment of revitalising these communities. She continued, "that is, to know why current Kuma is not deprived of a lively atmosphere as people are always visiting". A monthly gathering focusing on old people is one of her committed tasks. The objective of this gathering is to set up a social space focusing on old people, though not exclusively. She designs the programme for the monthly meeting at five different locations in the communities. In the first half of the programme, there are guest speakers or lecturers, such as a local police officer, public nurse, social worker and elderly care worker. The topics are varied among different speakers' professions who share a common thread of improving the quality of living for elderly people. The second half of the programme is more recreational, such as playing games, handicraft making or drawing. Overall, the space is interactive and full of lively talks between participants, NPO/*Suisha* members, and guests at each gathering. The participants also take simple health check-ups during the programme. Frequently, many of these check-ups show worrying results because the excitement of meeting each other raises their blood pressure. One participant mentioned that the importance of this gathering is the pre-planned meeting and time to gather with others, apart from the context of house visits in which community members usually meet.

Originally, the women who gathered from the different local women's collectives to start the food business collaboratively suggested the idea of a monthly gathering for the retired members. While doing activities such as fermentation, food processing, and farming at *Suisha*, the members started talking about their future. Indeed, *Suisha* members were happy about what they had made and accomplished, not only economic opportunities but also space for themselves and social companionship with others. However, this convivial environment is not ensured in the future when many members reach retirement age. With this notion, *Suisha* embarked on the plan of setting up space for themselves in the future, which became a space for *Suisha* members as well as other people in the communities, who are sharing the process of ageing.

Suisha's business practices take an approach that is based on hopeful and ambitious ethics that keep them in the business market while also caring for non-monetary benefits (Iwasaki & Miyaki, 2001). Their approach entails diverse ways of connecting with others and building interdependent relationships to improve everyday life, encompassing more than monetary benefits gained from the market. *Suisha* members attempt to sustain the ethical needs of "surviving well together", the central quality of what J. K. Gibson-Graham et al. (2013) call a community economy. In ageing and depopulating rural communities, they attended to the critical concerns of how to sustain a community. They did so by tapping into their situated knowledge of gendered everyday practices for their food business. They embodied economic practices with expanding caring relationships by blurring the binaries of capitalist and non-capitalist relations, men and women, and the boundaries of community membership. Moreover, along with the objectives as a business entity, they envisaged themselves as ageing entrepreneurial actors making nodes of connection with others, devising other forms and practices of surviving well together.

Constance: Uruguay, Red Nacional de Personas Mayores (REDAM)

In 2019, I was six months into my PhD, still working on settling into my topic and carving out a focus for myself at the intersection of ageing and environment, when I went to Uruguay on secondment. Soon after my arrival, I was introduced to the director of Uruguay's National Institute for the Elderly. Having shared with her my research interests, she suggested I attend the next meeting of REDAM, Uruguay's umbrella civil society space that brings together diverse organisations that work for/with/among older people.

Feeling timid with my out-of-practice Spanish, I was immediately welcomed. Curious about what a young foreigner was doing at a meeting of elderly Uruguayans, several members introduced themselves. I met a member of the local pensioners' association, recently retired and looking to stay politically active; the coordinator of a seniors' sports club; a nurse working in a care home; and a lifelong activist whose current focus was advocating for the full participation of older people in society. Though they each had different motivations for participating in REDAM, they were all invested in REDAM's objective to increase the political and social participation of older people in Uruguay.

It was my good fortune that this meeting happened to be the presentation of the network's working paper on "Key measures on ageing for the implementation and monitoring of the Sustainable Development Goals (SDGs)" (REDAM, 2019, p. 1). This seven-page document presented a clear critique of the SDGs as they currently stand, asking why the SDGs do not "address, either explicitly or implicitly, the rights of older people" (REDAM, 2019, p. 1). In the document, they go on to identify ways in which the network calls for further work to be done on ageing and environmental sustainability. I was still figuring out what FPE meant for me and for my work, and here I had come across feminist political ecology in practice.

I was invited to visit some of the member organisations that populate the network and dove deeper into the questions raised in their working paper. Over many cups of hot beverages, I learned more about the network's work to have older people taken seriously as subjects of rights in Uruguay. Despite the challenge of holding space for the diversity of interests present within the network, the REDAM has become a focal point for advancing the human rights of older people within Uruguay and across the region. REDAM members participated in the regional process that culminated in the Organization of American States' (OAS) Inter-American Convention on the Protection of the Human Rights of Older Persons (OAS, 2015). The network's members mobilised their wealth of knowledge about both domestic and international advocacy for the convention. Their important contributions to the process were recognised by other member states' civil society organisations for having been key to the signing and ratification of the convention (González Ballesteros et al., 2018). The REDAM's members have been able to draw on their everyday experiences of confronting ageism and imagining more full ways for older people to participate in society in order to navigate different local and international scales. Having gained a powerful international tool in the form of the OAS Convention, the REDAM's membership has also worked to bring these discourses back into their day-to-day through dissemination activities and by showing how the Convention can be utilised (Restaño, 2019), such as the above-mentioned critique of the SDGs.

I am especially grateful for the time I was able to spend with Luisa,[2] a lifelong activist who had worked alongside a handful of other REDAM members in the drafting of this critique of the SDGs from an ageing perspective. As a member of the network who has actively engaged with mobilising attention around the importance of participation of older people, she carried with her much of the institutional memory of the REDAM.

In recounting some of our conversations, I highlight the insights Luisa shared with me. When asked why they came to focus on the Sustainable Development Goals, she shared that she felt a responsibility to contribute

[2] Pseudonym.

to a world that could be a little more gentle than when she arrived in it. Luisa was also quick to point out the forgetfulness of younger generations. She underscored the irony in some of the sustainability discourse, that when people go looking for knowledge about how to tread more lightly on the planet, they often look elsewhere rather than with their grandparents and elders. Luisa shared this not in a resenting way, but to point to the pressure of time to be always looking forward, moving forward (which I would translate as capitalist productivist logics). She saw this pressure as meaning "there is no time to slow down and have the slower conversations with older generations". In addition to the importance of being included in a programme with societal relevance, Luisa felt she also had important contributions to make to the sustainability discussion. This was also reflected in the REDAM's document in reference to SDG 15—Life on Land: "As society, we must take into account the experiences of older people, their knowledge of the territory and of gentle forms of production with the Earth. Generational bridges should be built to allow for the transmission of these knowledges" (REDAM, 2019, p. 4).

Having had the chance to understand the process that led to the creation of the REDAM's SDG statement, I was very much looking forward to returning to Uruguay in early 2020 to see how the network was going to bring this document to life. Two factors coincided to make these plans impossible. First, after fifteen years of a progressive coalition government that achieved impressive and internationally recognised social gains, a conservative coalition won a fiercely contested election in the spring of 2019. This transition of government included the emptying out of many social institutions, including the National Institute for the Elderly which was the convenor of the REDAM. This process left REDAM a shell of its former self. It became clear that the work the network had been doing when I visited in early 2019 had, at least for the moment, been put to the side. The second factor was the COVID-19 pandemic. Despite these setbacks, Luisa and the REDAM's work has made important contributions, both nationally and regionally, to thinking about age and sustainability.

Understandings of Ageing from an FPE Perspective

These two experiences speak to distinct dimensions of ageing in vastly different contexts. Nanako brings out the interdependencies of care between older women, fungi and machines as well as a look into community economies from an ageing perspective. Constance touches on older people's agency and utilising civil society spaces to put forward their priorities at the intersection of sustainability and ageing. Despite the differences in these approaches to thinking about wellbeing in later life, putting them into dialogue has been fruitful for each of our understandings of ageing and has allowed us to carve out key insights at the intersection of ageing and FPE. In the following discussion, we draw out and elaborate on the broad themes of ageing and rurality, ageing and social/natural relations and community and economy.

Ageing and Rurality

Within studies on ageing, there is a clear divide between thinking about urban ageing and rural ageing in relation to demographic and social changes (Berry & Kirschner, 2013; World Health Organization, 2015). While insights into what distinguishes urban and rural ageing experiences can be helpful, this kind of binary thinking can also be limiting. In our above two studies, ageing experiences show that the urban and the rural settings are blurred; they are a web of different places, spaces, and environments where ageing experiences are materialised in everyday practices (Katz, 2018). In other words, the urban/rural divide obscures more than it illuminates; indeed, this picks up on wider literature within human geography that speaks of the rural as relational (Little, 2002). We also contribute to ageing geographies, which emphasise the ways older people interact with and are impacted by their environments (Andrews & Skinner, 2015; Skinner et al., 2015). As such, it may be more useful to relate ageing to individually unique experiences in everyday practices rather than to chronologically laid out demography (Aguirre Cuns & Scavino Solari, 2018; Katz, 2018).

Nanako's case may seem to conform to "rural ageing thinking", but *Suisha*'s business shows rural women's engagement with ageing by encouraging rural and urban interchange. Generally, the rural population is more drastically ageing than urban areas due to people leaving the countryside for education and work in addition to the older generation living much longer than in the past (Berry & Kirschner, 2013), which raises concerns about the sustainability in the everyday practices in the rural context (Ôno, 2008; Tamazato, 2009). Yet, rurality in the Japanese context can also represent attraction or nostalgia connected with the rural landscape (Tanaka, 2017). A rustic, mosaic landscape with diverse vegetation is a cohabiting space of humans and nature, in which people manage natural resources as a part of traditional rural livelihood practices. That type of landscape, and its embodied human and non-human relationships, evoke nostalgia and affectionate feelings associated with the image of "hometown" or *Kokyou/Furusato* in Japanese (Yukawa, 2017). That positive association is used in attempts at rural revitalization to attract more people to come and visit, and eventually settle down in the countryside (Love, 2007; Tanaka, 2017). Instead of performing rural ageing in a frame of the rural and urban dichotomy, *Suisha*'s business showed experiences of ageing together. *Suisha*'s business model places the traditional food practices of "hand-made" and "locally-produced" in the semi-mountainous remote area at the centre of their business. By doing so, they build and use their rurality as a business tool to sell their food products and attract more people to visit the shop and restaurant. In this way, challenges emerging from the rural context are addressed by non-binary rural and urban actors as a response to the sustainability concerns in the ageing and depopulating rural context.

In the Uruguayan experience, with almost half of the country's population living in the capital Montevideo, there exists a strong Montevideo/interior divide.[3] Though most of the REDAM members Constance came to know were from towns, the fact that they were from the interior positioned them differently—in a blurred, neither fully urban nor fully rural, experience. Many of these members faced the struggles of having to travel or relocate to access health and care services in larger centres, a common problem for rural ageing. And yet, through engaging in civil society spaces such as REDAM, many of these members' access to and participation in national and regional policy and activist spaces is more indicative of what is typically thought of as an urban experience of ageing. With regard to the REDAM's emphasis on sustainability, the network focuses on addressing climate issues as interconnected and brings together both urban and rural concerns. By virtue of being a national network with members representing each region and department of Uruguay, the REDAM has aimed to do justice to the specificity of its members' diverse

[3] With almost half of Uruguay's population living in Montevideo, the coastal capital city, the usual concentration of economic, cultural and political power that typically takes place in a capital is exacerbated in Uruguay. This has resulted in many disparities between the interior and the capital (Fernández Aguerre, 2017).

experiences while also articulating clear advocacy goals for the network as a whole. Their critique of the SDGs is a reflection of this. These examples add to the diversity of rural ageing experiences, underscoring the need for embodied and situated nuance.

Approach to Ageing and Social/Natural Relations

FPE's intersectional thinking (Rocheleau et al., 1996) offers a grounding to analyse gendered social differences that are embodied in environmental engagements (Harcourt & Nelson, 2015). FPE's relational thinking suggests a perspective that generation and age intersect with social differences, shaping power dynamics (Elmhirst et al., 2017; Nelson, 2017). Drawing on those strands of FPE, both our cases tease out the complexity of ageing people's relations and open up new opportunities to understand how to care for others in the past, present, and future, beyond the binary between human and nature (Nightingale, 2013). Contrastingly, the approach we have each taken has brought out different understandings of environment.

By illuminating care for relational ties in everyday practices, Nanako's case not only shows a space in which socionatural relationships emerge through *Suisha*'s business, but also questions the gaps in care between ageing rural women, miso, machines, and fungus. Ageing bodies experience lingering challenges in maintaining livelihoods with old fashioned machines and accessing social services such as elder care, even if *Suisha*'s business contributes to rural revitalization and the local economy. Thus, challenges come into play in shaping everyday ageing experiences in different places, spaces, and environments (Katz, 2018), yet these struggles also contribute to socionatural interdependencies. Still the gaps in care also demonstrate the need for additional services from social and political actors in the region.

With the more abstract entry point of discourses around the SDGs in the Uruguay case, Constance focused on how the REDAM's members' everyday practices of struggle. Their desires to be actively present in policy conversations about sustainability and care for the environment cross multiple temporalities. For some, the main motivation to participate touched upon hopes to protect and nurture places that had meaning for them, maintaining relationships to their pasts. Others were more motivated by the need to be heard, to be included in the debates, struggling against the invisibilising of older people. As mentioned, Lucia identified a responsibility to future generations as a motivation for her. Also, understandings of mutuality were prevalent in Uruguay, with participants speaking to the idea of treating the environment gently and it too will treat you gently.

Drawing on FPE's approach to social and natural relations allows us to see that possibilities for continuing everyday life emerge from interactions of ageing others and different places and environments. Though not explored in depth in our narratives above, we found it interesting that people of varying ages participated in each of the groups. Some of the younger participants in both spaces spoke to the importance of investing in spaces for them to continue to be engaged and active in their own old age- an investment in a future temporality of sorts. While we started questioning the division in urban and rural ageing, our attempts of bringing ageing to FPE can shift the focus to tune in to other divisions between humans and nature, young and old, and other dichotomies alike.

Community Economies and Post-capitalist Perspectives on Ageing

Drawing on Jean-Luc Nancy's "being-in-common" (1991), post-capitalist community economies' understanding of care offers insights into the interdependency of various economic beings and how individuals become communities in order to survive well together (Gibson-Graham, 2006). As shown in the Japanese case, ageing experiences embodied by rural women involve care for other humans and non-humans. *Suisha's* survivall relies on caring for others and themselves in the ageing process; that is, not just ageing well individually, but also surviving well together with ageing others by appropriating profits gained from their food business for other-than-capitalist wellbeing. These practices that go beyond the binary thinking of capitalist and non-capitalist framing rest on care for others, and themselves in the context of an uncertain future cannot be static or decisive. By unravelling the process of how rural women make different decisions in everyday practices, Nanako's case emphasised that ageing experiences are relational and constituted together with others based on their ethics of surviving well together. The ethics of care for others also embraces the sensibility of ageing with others, contributing to a collective experience and construction of ageing. Thus, post-capitalist thinking can highlight how actors are not framed by limited definitions such as women and men, human and nature, but rather, are seen as emerging economic beings (Nancy, 1991; Gibson-Graham, 2006).

Ageing beings are caring for others in both cases, constructing communities based on interdependent relationships through pursuing *Suisha*'s local business in Japan and climate justice from the elderly perspective in Uruguay. They produce and reproduce interdependent relationships, the foundation of community economies, rather than championing care as outside of profit making or denying capitalists' way of wellbeing. This relational, non-binary thinking examines how to "survive well" together with humans and non-humans by balancing business practices and wellbeing while caring for the differently interdependent economic relationships (Gibson-Graham et al., 2013, p. 21).

Rupturist Gerontology

In analysing the Uruguayan case, rupturist gerontology (Piña Morán & Gómez Urrutia, 2019) helps to understand the link between age and gender by analysing social roles. Rupturist gerontologists and feminist writers both reject the framing of old age as a period "without activity" (ibid.; p. 14). This approach breaks with the idea of the elderly as "subjects of care", overcoming the dichotomy of receivers and givers. Through this lens, care becomes a relational feature that informs our interactions with others and with life in its diversity. The complex motivations of the members of REDAM speak to this relational, sometimes contradictory, complexity. The clear demand to be seen and heard is an attempt to counter the invisibilisation of older people. Their choice as a collective to focus on an issue such as environmental sustainability speaks to their willingness to invest in future wellbeing they may not be present to enjoy. These are both motivations that are focused on building relations. REDAM members could thus be seen as demanding more care and demanding more contexts in which they can offer care. This relational approach to care, essential within FPE and articulated strongly within rupturist gerontology, places older people in relation with others, making decisions collectively and imagining new realities, recognising the knowledges, perspectives, shortcomings, experiences and limitations that are part and parcel of older people's livelihoods.

Concluding Thoughts

In this chapter, we present a case for including ageing in the current contours of FPE research. In demonstrating the value of such an analysis, we have brought two distinct ageing experiences—*Suisha*, the women's local business in Japan, and REDAM, a national network for older people in Uruguay—into the conversation. We highlighted that ageing is neither fixed nor binary but is about the embodied experiences and agency of older people. Our distinct theoretical groundings and our different interpretations of FPE allowed us to learn from each other's case and deepen our understanding of our own. These threads emerged at the intersection of ageing and FPE, enriching the understanding of ageing experiences.

In both cases, we draw on ethics of care (Puig de la Bellacasa, 2017) in thinking about older people and the ageing process. The *Suisha* case shows care for *kôji* and other people in the past, present, and future, while REDAM's case shows care for future generations by ageing people claiming their own presence. Understandings of socionatural relations have also helped us explore more deeply the layered relations in which ageing takes place.

Age is often overlooked or under-examined as both a dimension of power and as a bio-social process within FPE-inspired work. As both cases show, age is a central factor within embodied care practices and enhances the interdependency in socionatural relationships that support wellbeing. With ageing itself being a series of ongoing relational changes, we have made the case for what can be gained at the intersection of ageing and environment in the hopes that others also take up questions of age and intergenerational wellbeing within research on embodiment, socionatural relationality and everyday practices of care.

Acknowledgements We would like to express our deep gratitude to Ana Agostino for her ongoing support in bringing these ideas together. Also, many thanks to Rebecca Elmhirst and Wendy Harcourt for devoting their time and energies to reviewing our manuscript and providing us with valuable suggestions.

Funding: This chapter was funded by the Wellbeing Ecology Gender and cOmmunities Innovation Training Network (WEGO-ITN) funded by the European Union's Horizon 2020 research and innovation programme under the Marie Sklodowska-Curie grant agreement No. 764908-WEGO 2018-2021.

References

Aguirre Cuns, R., & Scavino Solari, S. (2018). *Vejeces de las mujeres. Desafíos para la igualdad de género y la justicia social en Uruguay*. Doble Click Editoras.

Andrews, G. J., & Skinner, M. W. (2015). Aging and older people, geography of. In J. D. Wright (Ed.), *International encyclopedia of the social & behavioral sciences* (2nd ed., pp. 407–412). Elsevier.

Berry, E. H., & Kirschner, A. (2013). Demography of rural aging. *Rural Gerontology* (pp. 17–36). Routledge.

Calasanti, T., Slevin, K. F., & King, N. (2006). Ageism and feminist: From "Et Cetera" et Center. *National Women's Studies Association Journal, 18*(1), 13–30.

Calasanti, T. (2008). A feminist confronts ageism. *Journal of Aging Studies, 22*, 152–157. https://doi.org/10.1016/j.jaging.2007.12.009

Dillaway, H., & Byrnes, M. (2009). Reconsidering successful aging: A call for renewed and expanded academic critique and conceptualizations. *Journal of Applied Gerontology, 29*(2), 109–130. https://doi.org/10.1177/0733464809333882

Elmhirst, R., Siscawati, M., Basnett, B. S., & Ekowati, D. (2017). Gender and generation in engagements with oil palm in East Kalimantan, Indonesia: Insights from feminist political ecology. *The Journal of Peasant Studies, 44*(6), 1135–1157. https://doi.org/10.1080/03066150.2017.1337002

Fernández Aguerre, T. (2017). Brecha geográfica Montevideo-interior en los logros educativos de Uruguay (1996–2015). *tópos, 9*, 14-20.

Gibson-Graham, J. K. (2006). *A postcapitalist politics*. University of Minnesota Press.

Gibson-Graham, J. K., Cameron, J., & Healy, S. (2013). *Take back the economy: An ethical guide for transforming our communities*. University of Minnesota Press.

González Ballesteros, L. M., López Zuluaga, S., Ortíz Hoyos, A. M., Díaz Oramas, A., & Bustamante Torres, M. (2018). *El papel de la sociedad civil en el proceso de firma y ratificación de la Convención Interamericana sobre la Protección de los Derechos Humanos de las Personas Mayores (CIPDHPM)* (Civil society report). HelpAge International and Fundación Saldarriaga Concha. Retrieved from https://www.helpage.org/silo/files/resumen-ejecutivo-estudio-sociedad-civil-y-ratificacion-de-la-convencion-interamericana.pdf

Harcourt, W., & Nelson, I. L. (2015). *Practising feminist political ecologies: Moving beyond the 'green economy.'* Bloomsbury Publishing.

Iwasaki, Y., & Miyaki, M. (2001). *Seikôsurunôsonjoseikigyô: shigotochiikijibunzukuri*. Ie-no-hikari-kyoukai.

Katz, S. (2018). *Ageing in everyday life: Materialities and embodiments*. Bristol University Press.

Katz, S., & Calasanti, T. (2015). Critical perspectives on successful aging: does it "appeal more than it illuminates"? *Gerontologist, 55*, 26–33. https://doi.org/10.1093/geront/gnu027

Little, J. (2002). Gender and rural geography: Identity, sexuality and power in the countryside (1st ed.). Routledge. https://doi.org/10.4324/9781315838717

Love, B. (2007). Fraught fieldsites: Studying community decline and heritage food revival in rural Japan. *Critical Asian Studies, 39*(4), 541–559. https://doi.org/10.1080/14672710701686034

Nightingale, A. (2013). Fishing for nature: The politics of subjectivity and emotion in Scottish inshore fisheries management. *Environment and Planning A, 45*(10), 2362–2378. https://doi.org/10.1068/a45340

Nightingale, A. (2017). *Environment and gender*. Wiley.

Nightingale, A. J. (2018). The socioenvironmental state: Political authority, subjects, and transformative socionatural change in an uncertain world. *Environment and Planning E: Nature and Space, 1*(4), 688–711. https://doi.org/10.1177/2514848618816467

Nancy, J. L. (1991). Of being-in-common. In Theory Collective (Ed.), *Community at loose ends* (pp. 1–12). University of Minnesota Press.

Nelson, I. L. (2017). Interspecies care and aging in a gorilla 2.0 world. *Geoforum, 79*, 144–152. https://doi.org/10.1016/j.geoforum.2016.02.007

Organization of American States. (2015). *Inter-American convention on protecting the human rights of older persons* (A-70). Retrieved from http://www.oas.org/en/sla/dil/inter_american_treaties_A-70_human_rights_older_persons.asp

Ôno, A. (2008). Kyôto Shinbun Shuppan Sentâ. Kyôto shinbun kikaku kigyô.

Piña Morán, M., & Gómez Urrutia, V. (2019). Envejecimiento y género: Reconstruyendo los roles sociales de las personas mayores en los cuidados. *Revista Rupturas, 9*(2), 23–38.

Puig de la Bellacasa, M. (2017). *Matters of care: Speculative ethics in more than human worlds*. University of Minnesota Press.

REDAM. (2019). *Documento de trabajo de la Red de organizaciones de personas mayores de Uruguay (REDAM) referente a: "Medidas clave sobre envejecimiento para la implementación y seguimiento de los Objetivos de Desarrollo Sostenible"*, Neustra Voz +60, 145. Retrieved from: https://issuu.com/publibook13/docs/nuestra_voz_n__145_para_web

Restaño, Á. (2019). *Presentación REDAM por la Reunión de Expertos "Medidas clave sobre envejecimiento para la implementación y seguimiento de los Objetivos de Desarrollo Sostenible"* (Civil society report). CEPAL Meeting. https://www.cepal.org/es/eventos/reunion-expertos-medidas-clave-envejecimiento-la-implementacion-seguimiento-objetivos

Rhee, J. E. (2020). *Decolonial feminist research: Haunting, rememory and mothers*. Routledge.
Rice, C., Chandler, E., Rinaldi, J., Changfoot, N., Liddiard, K., Mykitiuk, R., & Mündel, I. (2017). Imagining disability futurities. *Hypatia, 32*(2), 213–229. https://doi.org/10.1111/hypa.12321
Rocheleau, D., Thomas-Slayter, B., & Wangari, E. (1996). Gender and environment: A feminist political ecology perspective. In D. Rocheleau, B. Thomas-Slayter, & E. Wangari (Eds.), *Feminist political ecology: Global issues and local experiences* (pp. 3–23). Routledge.
Sandberg, L. (2015). Towards a happy ending? Positive ageing, heteronormativity and un/happy intimacies. *lambda nordica, 20*(4), 19–44.
Sandberg, L. J., & Marshall, B. L. (2017). Queering aging futures. *Societies, 7*(3), 21. https://doi.org/10.3390/soc7030021
Sato, C., & Alarcón, S. (2019). Toward a postcapitalist feminist political ecology' approach to the commons and commoning. *International Journal of the Commons, 13*, 36–61. https://doi.org/10.18352/ijc.933
Singh, N. (2017). Becoming a commoner: The commons as sites for affective socio-nature encounters and co-becomings. *Ephemera: Theory & Politics in Organization, 17*(4), 751–776.
Skinner, M. W., Cloutier, D., & Andrews, G. J. (2015). Geographies of ageing: Progress and possibilities after two decades of change. *Progress in Human Geography, 39*(6), 776–799. https://doi.org/10.1177/0309132514558444
Tamazato, E. (2009). *Shûrakugankaikawokoete: Shûrakusaiseie Kôchi karahassin*. Fukurô shuppan.
Tanaka, S. (2017). *Toshinoakogaresansonnotomadoi: kyōtofumiyamachōtoiuyume*. Kōyōshobō.
Tronto, J. C. (1993). *Moral boundaries: A political argument for an ethic of care* (1st ed.). Routledge.
Vasquez, R. (2020). *Vistas of modernity: Decolonial aesthetics and the end of the contemporary*. Mondriaan Fund.
World Health Organization. (2002). *Active ageing: A policy framework*. WHO.
World Health Organization. (2015). *World report on ageing and health*. Retrieved from http://apps.who.int/iris/bitstream/10665/186463/1/9789240694811_eng.pdf
World Health Organization. (2020). *Decade of healthy ageing: Baseline report* (9240017909). Retrieved from https://www.who.int/publications/i/item/9789240017900
Yukawa, M. (2017). 'Nihonnogenfuukei'tobunkanationalism: Kokyôwoshouhisurutoshi, Kokyowosouzousurunousanson. In S. Tanaka (Ed.), *Toshinoakogaresansonnotomadoi: kyōtofumiyamachōtoiuyume* (pp. 239–256). Kōyōshobō.

CHAPTER 6

More-Than-Human Co-becomings: The Interdependencies of Water, Embodied Subjectivities and Ethics

Nick Bourguignon, Irene Leonardelli, Enid Still, Ingrid L. Nelson, and Andrea J. Nightingale

Introduction

In a small drought-prone rural village in Maharashtra, Pravah,[1] India, farmers irrigate their farms using the wastewater of the city of Pune. Water becomes multiple and troubled as it is used, embodied, and experienced in different ways. Water(s) and women farmers co-become through their varied relationships to one another. Similarly, the journey of the Tagus River through Spain and Portugal tells a story of multiple, troubling waters interwoven with landscapes, people, histories and more-than-human co-becomings. The Tagus as an entity is shaped by multiple

[1] All names of people and dwelling places are pseudonymised, names of rivers and infrastructure are not changed.

N. Bourguignon
University of Barcelona, Barcelona, Spain
e-mail: nico.bourguignon@gmail.com

W. Harcourt et al. (eds.), *Contours of Feminist Political Ecology*, Gender, Development and Social Change,
https://doi.org/10.1007/978-3-031-20928-4_6

human interests—irrigation, human consumption, international relations—but also shapes human communities within and beyond its basin. These stories raise contradictions and ethical concerns that highlight the importance of thinking through more-than-human interdependencies to understand how water- and hydro-scapes emerge.

Whether a village served by a wastewater scheme in Maharashtra, India, or an inter-basin water transfer infrastructure in Spain, hydraulic schemes and infrastructures create webs of relations within and across multiple scales. This inspires us to grapple with more-than-human concerns. The term "more-than-human" situates humans within the sticky webs of relations, making humans interdependent with other beings and materials (Barad, 2007; Haraway, 2003; Isaacs & Otruba 2019). We focus on the relationality of the more-than-human to consider the ethical contradictions that flow of power within these webs engender. These relations can be life-giving or denying (Bawaka Country et al., 2013; Singh, 2013; Zwarteveen & Boelens, 2014), create socionatural difference and inequalities (Ahmed & Zwarteveen, 2012; Harris, 2006; Nightingale, 2006; Sultana, 2009), and permeate the co-constitution of embodied subjectivities (Nightingale, 2006, 2011; Singh, 2013). Particularly, we focus on water and the ways it co-constitutes uneven relations between and within different human beings, other beings and materials across multiple times and scales (Neimanis, 2012). In doing so, we recognise that our attempt to grapple with the more-than-human is always situated, partial and shaped by our positionalities.

I. Leonardelli (✉)
IHE Delft Institute for Water Education, Delft, The Netherlands
e-mail: i.leonardelli@un-ihe.org

E. Still
University of Passau, Passau, Germany
e-mail: Enid.Still@Uni-Passau.De

I. L. Nelson
University of Vermont, Burlington, VT, USA
e-mail: prof.ing.nel@gmail.com

A. J. Nightingale
University of Oslo, Oslo, Norway
e-mail: andrea.nightingale@sosgeo.uio.no

In this chapter, Irene Leonardelli presents the section dedicated to Pravah in the second person, recognising herself as a researcher generating data together with a translator as well as with local researchers-activists and informants in the village. Nick Bourguignon presents the section dedicated to the Tagus River in the first person, recounting his experiences and knowledge of the river. Enid Still engages with these narratives philosophically by thinking through the ethical implications of more-than-human interdependencies. Ingrid Nelson and Andrea Nightingale have been co-thinkers through the entire process, helping to find a narrative thread, navigate the wide literatures on socionatures and reflect on feminist political ecology (FPE) praxis. Together, we aim to think of waters through different lenses and in different geographies, with specific histories and socialities, while remaining aware that our human histories and privileges shape these representations of waters. This movement between different positions of the self in relation to water and the more-than-human is intentional, as we feel it reflects the way water is unbound, percolates through bodies and soils, shapes both alliances and conflicts between different actors, such as the state, community organisations and farmers.

Framing more-than-human interdependencies within FPE means *starting* from an understanding of relationality rather than the experience of distinct individuals (Nightingale, 2011; Rowe, 2005). Exploring this relationality as entangled, multiple and situated has enabled feminist political ecologists to contribute to debates on how subjectivities are co-constituted with the more-than-human (Nightingale, 2006). Andrea Nightingale (2013) demonstrates that Scottish fishermen's emotional and political subjectivities are co-constituted by the sea; it can give or take life, it can create a sense of purpose and bend political will, and it can create or destroy possibilities for fishermen. Secondly, intersectional inquiry into how power operates within these entanglements has enabled nuanced accounts of how gender, caste, class, age, seniority, marital positions and ethnicity are differently implicated in the community management of ecological resources (Elmhirst et al., 2017; Nyantakyi-Frimpong, 2019). Furthermore, FPE questions how socionaturally constituted power dynamics refract upon hierarchical social relations and boundaries, as well as wider systems of power such as colonialism (Nelson, 2017; Nightingale, 2011).

For this chapter, we draw on this work in FPE and related disciplines, focusing on the development of the concept of more-than-human interdependence in an attempt to trouble essentialising narratives that may be read into such complex and situated relationships (Barad, 2007; Bawaka Country et al., 2013; Haraway, 2003; Nelson, 2017; Nightingale, 2013; Rose, 2012). In doing so, we remain aware that writing about the more-than-human is a socially mediated process, despite the multiple forms of more-than-human agency that go beyond the immediate human experience.

The next section begins with the case of Pravah, where a wastewater reuse scheme allows farmers to farm throughout the year, cultivating different crops both for household consumption and to sell at the market. However, the wastewater that is used for irrigation percolates into the village's shallow aquifer contaminating the existing, scarce, water. Women, who primarily bear responsibility for water, have to navigate different levels of contamination at water sources and adjust their practices accordingly. The concept of "waterscape" flags an ontological definition of water where power is enrolled in co-constitutive representations and materialities of water (Baviskar, 2007). Next, we journey with Nick's account, where the river's ebbs and flows constitute fuel and food that sustains the lives of many beings, as well as embodied memories of co-becoming with the multiple existences of "the Tagus". Bubbling forth in Eastern Spain, the Tagus cuts through canyons and pine forests, meanders past Toledo and across plains and *dehesas*, joining the Atlantic Ocean in Lisbon. Yet the waters also join with the Mediterranean through the Tagus-Segura Interbasin Water Transfer. Wastewater from Madrid's 6.6 million inhabitants joins the Tagus with wafts of ammonia, to bubble through Toledo. The concept of "hydrosocial territory" helps explore the multiple scales of contested socionatural materializations of spatially bound networks of the river (Boelens et al., 2016). Finally, we discuss the situated knowledges and ethical implications that emerge from these ethnographic explorations and the way they illuminate interdependencies in particular ways that can inform processes of co-becoming with water. Each of the following sections discuss how such interdependencies are lived and yet often obscured, and how thinking with the politics of interdependency can enable us to narrate the situated complexities of their emergence as a process of co-becoming.

Gender-Water Intra-Action in a Wastewaterscape in Maharashtra, India

> The main public well in the village is called Sakarbai [sugar lady] because its water used to be pure, fresh and sweet. (…) Water was very scarce outside the monsoon season, but it was so good that it used to taste like coconut water. She [Sakarbai] used to give us the best water even in summer, when we used to climb down the well to get whatever water we could find. But after we started using wastewater for irrigation, she [Sakarbai] was really badly affected. It's not her fault, but her water became disgusting, so we stopped drinking it.
>
> Interview with Sonali, 28 February 2020

Sonali, a seventy-year-old woman belonging to the Maratha caste -the upper caste in the village of Pravah- and who worked her entire life as a farmer, is talking while sitting on the floor of her house. Sakarbai is the name of the well where she used to fetch water for drinking and other domestic purposes since she moved to Pravah as a just-married child, at the age of six. Around 2009, Sakarbai started changing. It happened as more farmers in Pravah started buying wastewater to irrigate, taking advantage of a wastewater reuse system called the Purandar Lift Irrigation Scheme. This was designed by the Government of Maharashtra in the early 2000s to address water scarcity in 60 rural villages located in the Purandar sub-district, southeast of Pune. Wastewater is pumped up from the Mula-Mutha river, which flows through Pune collecting untreated water from the urban sewer system and industrial effluents, manufacturing industries, construction sites, automotive garages and hospitals (Jagtap & Manivanan, 2019). Farmers can buy wastewater outside the monsoon season, usually from January until June when their wells are empty. Along with traditional rainfed crops such as *bajra* (pearl millet) and *jowar* (sorghum), they use wastewater to irrigate different non-traditional crops, particularly flowers, to sell at markets in nearby towns and in Pune. Wastewater reaches Pravah through a system of pump houses and pipelines; farmers store it in private water ponds (Leonardelli et al., 2022). From these ponds, wastewater percolates into the shallow aquifer of the village, contaminating the existing, albeit scarce, groundwater. This way, all wells (those used for irrigation as well as Pravah's former drinking water well Sakarbai) get recharged with (at least partly) contaminated water.

As the Purandar Lift Irrigation Scheme changes water flows, a wastewaterscape (Karpouzoglou & Zimmer, 2016) emerges. This entails new material-discursive relations between different humans and waters. Wastewater opens up possibilities (e.g., for irrigating and thus farming differently) and closes down others (e.g., drinking pure and sweet groundwater), re-articulating more-than-human relations in troubling ways that raise ethical questions and concerns. Here we unfold some of these re-articulations focusing on the interdependencies between women farmers[2] and water, thereby troubling boundaries that signify what is considered an ethical encounter. We make this choice because in Pravah, like throughout Maharashtra and other parts of rural India, fetching water, cooking, cleaning, washing and bathing children are tasks performed mostly by women. Women also perform much of the everyday work in the farm, work which is profoundly mediated by water (Krishna & Kulkarni, 2019). The focus on women farmers also stems from our ethico-political commitment to complicate dominant representations, challenging the processes through which women farmers' experiences, knowledges and practices are marginalised and/or silenced, especially those of women farmers of Scheduled Castes,[3] landless women, single women and widows (Bhat, 2016). Indeed, the conceptualization of "farmers" in India most often reproduces the imaginary of farmers as male landowners (Agarwal, 2003; Padhi, 2012; Still, 2022).

Feminist scholars studying water in rural, agrarian contexts have shown how gender and water are intimately interwoven (Ahmed & Zwarteveen, 2012; Bossenbroek & Zwarteveen, 2018; Harris, 2006; Mehta, 2014; Sultana, 2009). Their interdependence is simultaneously material, symbolic and discursive, with "gender" and "water" (and all that emerges in their relation for instance farming) continuously co-constituting and re-defining one another (Bossenbroek & Zwarteveen, 2018). Re-allocations of water, for instance, those fostered by neoliberal processes of development, imply changes in gender labour relations, roles

[2] We use the category of "women farmers" to refer to people who, during our fieldwork, identified themselves as such, though sticking to a definition of gender as a fluid "performative accomplishment" (Butler, 1990) shaping a myriad of different subjectivities.

[3] "Scheduled Caste" is a politically imposed category that encompasses all castes considered to be outside the caste or varna system, and who are therefore systemically oppressed and socio-economically disadvantaged. It is a recognised term in the Constitution of India and is used to enable reservation status for those from scheduled castes, to enable their representation in social and political life (see Gnana, 2018).

and responsibilities, thus also re-defining gendered embodied subjectivities (Harris, 2006). At the same time, experiences such as headaches and backaches from hauling heavy loads of water, as well as the health implications arising from consuming unsafe water, illustrate how embodied gendered subjectivities are spatially produced through everyday dealings with water (Sultana, 2009).

These studies have pointed out how everyday dealings with water significantly shape gendered embodied subjectivities. Yet what we aim to emphasise here is that how water behaves in and through specific landscapes (seeping, percolating, overflowing, evaporating) and what it transports (algae, sediments, organic matters, contaminants) also play a role in co-constituting specific gendered embodied subjectivities.

In order to reflect these fluid sensibilities, we draw inspiration from post-human and Science and Technology Studies (STS) feminist scholars, as they cultivate sensitivity for what more-than-humans, and water particularly, trigger and afford in socionatural relations (Barad, 2007; Neimanis, 2017; Pickering, 2009). Particularly, Barad (2007)'s concept of intra-action helps focus the analysis on how human bodies and more-than-human bodies co-constitute one another simultaneously, discursively and materially, meaning they do not exist as separate, discrete entities. This conceptualization helps us decentre the idea that humans dominate matter to instead pay more attention to the specific behaviours that different humans and more-than-humans, including water, afford as they relate. This becomes useful when unpacking how gendered subjectivities embodied by women farmers of different castes and classes and water co-become in the wastewaterscape of Pravah.

The Multiple Waters of Pravah

Around 2011, soon after the wastewater reuse scheme started functioning, Sonali and all other women farmers across caste and class noticed that the colour of Sakarbai water had become more and more turbid; its taste was more bitter, and unpleasant. At the same time, people fell sick with stomach diseases, which they related to the water they were drinking. These sensorial and embodied experiences laid a foundation for women farmers across castes to re-signify what water is: how and for which purposes it could be used. Though they appreciated the abundance of water in Sakarbai, as they no longer had to depend on government-sponsored water trucks or walk long distances to fetch water during the driest months of the year, they deemed Sakarbai water as unclean, smelly, "full of chemicals" and "dangerous". "*It's not her fault*" said Sonali: recognising how this exogenous water flow damaged the local "sweet" and "pure" water of their aquifer. For a few years, the Gram Panchayat [village council], in consultation with the local doctor, added a "medicine" to Sakarbai water to disinfect it, to heal it, before using it for drinking and cooking.

Then around 2016, the Gram Panchayat decided to instal a water vending machine (commonly known as a water ATM) behind the main square of the village. This technology purifies Sakarbai's water through a reverse osmosis system to make it safe to drink. Most women farmers now buy this filtered water for drinking; those who can afford it also use it for cooking- mostly women farmers from the Maratha (upper) caste. Other women (mostly those belonging to Scheduled Castes) and the least well-off farmers of the Maratha caste still go to Sakarabi to fetch water for cooking. Only women farmers belonging to a less well-off caste (a Non-Scheduled herding caste) and residing about 2 km away from the main village do not buy filtered water for drinking but disinfect the one they find in their irrigation well. For them, filtered water is too expensive and too long a walk.

For washing clothes, cleaning and bathing, women of all castes use the water they get from the taps installed outside of almost every house of the main village. Tap water is pumped from two public wells located close to several wastewater ponds, transported through a closed pipeline to a water tank, where it is disinfected, and then distributed to the taps. Women farmers told us that this water remains highly contaminated: they would never use it for drinking or cooking, not even after disinfecting it or boiling it themselves. Sometimes it smells badly, especially during the driest months when they buy the greatest quantities of wastewater. Moreover, women farmers of all castes often complain about irritations and rashes on their arms and legs, as well as about hair loss as they use tap water for bathing. Women farmers of the herding caste residing far away from the main village do not have access to tap water as the pipelines connected to the water tank do not reach as far. Since their shelters are located next to the wells they use for irrigating, they pump water directly and use it for domestic uses. They say that well water is pure enough as it percolates from the irrigation ponds where wastewater is initially stored, and in the process is purified by the soil.

In fact, in the narratives of women farmers across castes, the materiality of wastewater changes as it percolates through the soil of Pravah, mixing with groundwater: the longer it percolates, the more it purifies and becomes somehow less exogenous, and therefore also less "bad" and "dangerous" (see Leonardelli & Tozzi, forthcoming). The wastewater stored in the irrigation ponds is deemed the most contaminated. Women farmers know this as they use it across the farm: it smells badly, it contains algae and looks foamy and turbid, especially when it is just delivered. They told us that animals (cows and goats) often get sick if they drink directly from those ponds.

As women farmers navigate and make sense of water quality through the landscape of Pravah, water becomes multiple waters. At different sources, they get water for different purposes; they store them separately and treat them differently. Significantly, while everyone in the village at least partly adjusts their water practices, more well-off farmers (mostly belonging to the upper Maratha caste) have better means to deal with the consequences of this contamination: for instance, buying filtered water for multiple purposes and not just for drinking, accessing medicines and health services if necessary (see also Mehta & Karpouzoglou, 2015).

At the same time, because wastewater percolates in the shallow aquifer, it brings new irrigation and farming opportunities to all farmers of Pravah, not only to those who can afford to buy wastewater for irrigation. Well-off farmers belonging to upper castes can easily access the benefits of the Purandar Lift Irrigation Scheme; they can afford the cost of wastewater as well as the agricultural inputs required to engage in irrigated farming, cultivating diverse crops throughout the year (Leonardelli et al., 2022). Yet less well-off farmers (who are usually of Scheduled Castes and the herding caste) also find more water in their wells: some of them are able to farm on larger plots of land, including commercial crops like flowers. This way, as wastewater percolates through the landscape of Pravah, it benefits farmers across castes, even those that do not have the means to access the irrigation scheme directly.

These new farming opportunities mean both new responsibilities and work burdens for women across castes (Leonardelli et al., 2022). While they play an increasingly important role in deciding what to cultivate and how to organise the work, commercial cash crops such as flowers require a lot of care and effort. Women farmers across castes are increasingly involved in irrigation and in spreading pesticides- tasks that have historically been part of the male domain. Yet, wastewater fosters the growth of unwanted weeds, and thus long hours of strenuous weeding work. Women farmers need to carefully supervise the irrigation process to clear the drips when they are plugged by sediments and algae and wash them with an acid lotion after every cropping cycle (see also Leonardelli & Tozzi, forthcoming). This sheds light on how what wastewater transports and what it is made of also matters in re-articulating everyday work in the farm, and thus also in shaping embodied subjectivities.

Through their intra-action, (waste)water and women farmers co-become in Pravah, in ways that sometimes reinforce and sometimes go beyond caste differences (Nightingale, 2011; see also Leonardelli et al., 2022). In this regard, the relations that enact the wastewaterscape of Pravah are troubling and ambiguous: as wastewater flows from the city to rural areas, it carries particles that co-constitute human and more-than-human bodies. It allows farmers across castes to cultivate and sell more crops throughout the year, but it also pollutes the aquifer, badly affecting the health and well-being of more-than-human bodies, including people's bodies, water, soil, animals (Leonardelli & Tozzi, forthcoming). While farmers, including women farmers across castes, are generally satisfied with being able to sell crops throughout the year and to diversify their livelihood, they -and we, with them- remain entangled in the different "goods" and "bads" at play from using untreated wastewater for irrigation (Abrahamsson et al., 2015). The ethical slipperiness of these more-than-human entanglements enables reflection on the nature of ethics when conceptualised within processes of more-than-human co-becoming- a conceptualisation we grapple with in the final section.

Water Across Time, Space, Basins and Subjects—The Tagus River

Recognising individual human attachment to place, territory or country infuses the political into debates within and across more-than-human entities that pulse through multi-scalar spaces (Bawaka Country et al., 2013). It is also a way of thinking through the troubles of one's positionality in relation to spaces that are always co-constituted by the intra-action between oneself, other humans and more-than-humans (Barad, 2007). Storytelling of the more-than-human (Multispecies Editing Collective, 2017; Tsing, 2015) has the potential to narrate complex stories about the entanglements of life (Fenske & Norkunas, 2017). I therefore tell my own story of interdependence with a river that has been present in my life and is the subject of my own research—the Tagus River.

Visiting stretches of the river reveals different realities that a dominant discourse may hide. Where a political ecology of water (Baviskar, 2007; Boelens et al., 2016; Swyngedouw; 1999) helps think of large and complex hydrosocial territories, feminist contributions (Harris 2006; Neimanis, 2012; Sultana, 2011) emphasise the situatedness, subjectivities and relations of human bodies vis-a-vis more-than-humans, often articulated through the concept of waterscapes. By articulating this story from my situated feminist perspective, I see the Tagus River as a hydrosocial territory and explore ethical dilemmas within it. Hydrosocial territoriality helps me think of the divergent discourses that (re)produce material relations and subjects within a (dominant) political order, which incorporates multiple territorialities across scales, different actors—particularly irrigation communities[4]—and where dominant modes reconfigure material relations and subjects through particular water truths and knowledge claims (Harris 2006; Boelens et al., 2016). The frictions that emerge when non-dominant territories are pulled into a political order are captured in how inhabitants of the Tagus basin—myself included—narrate, live and experience the river.

The Divergent Flow(s) of the Tagus

The Tagus emerges in a pine grove at 1,600 metres above sea level in Eastern Spain. It leaps westward, gaining speed, until it flows through deep canyons lined with willows and poplars and eagles fly above. This early stretch has human visitors camping and hiking alongside the clear green water, as my family has done, connecting to the riverine landscape produced over geological time. Historically, loggers navigated pine logs along the canyon walls down to the city of Toledo, where they became embedded in buildings and cities.

The river encounters its first human obstacles in the reservoirs of Entrepeñas, Buendía and Bolarque. Since 1979, some Tagus waters flow artificially southeast, across the plains of Castille la Mancha through 286 kilometres of canals and pipes, into the Mundo River and then into the

[4] Irrigation communities are historic as well as contemporary water user associations that use historic and existing irrigation schemes. They are regulated by law (2001 Spanish national hydrological plan) and receive public water concessions from river basin authorities (Hernández-Mora et al., 2014).

Segura River, watering the historical and expanding irrigation fields of the Spanish Levant (Morote et al., 2020).

Exiting its original canyons, the river meanders down the plains of Guadalajara, with less flow and energy due to the syphoning off to Spain's levant. It reaches Aranjuez, where irrigation communities have cultivated strawberries for centuries (Moreno, 1980), and a few kilometres further downstream the Jarama River joins it. The Jarama is artificially swollen, receiving water transfers from other tributaries to slake the thirst of Madrid's 6 million inhabitants and industries, bringing with it waste. Urban residents of Madrid are physically embodied in the waste that flows down the Jarama into the Tagus, even if awareness of this finishes at the flush of a toilet. As a child, I played in and fell ill from this water that had strong tangs of ammonia.

The polluted river reaches the Castrejón reservoir, before it is channelled along the Castrejón canal. While the canal waters fields, the river itself pulses agonisingly along, reaching the mediaeval bridge of Montalbán. Standing on this bridge, the stillness in the air is pierced by shrieks from a nearby slaughterhouse, and the polluted river turns from brown to red. Amid the stench, storks retrieve unidentifiable things from the water. Witnessing this stretch, through what is seen and smelt, the river is embodied within myself. I react, rejecting these senses. They inform an ambivalent subjectivity, where I directly experience and feel attachment to the river while simultaneously feeling disconnected to its rural stretches by virtue of my identification as both urban and foreign. This then informs a personal politics that is inspired by how the river *is* in other stretches—clear, flowing, rich. Following Barad's (2007) intra-action, this subjectivity is an example of material and discursive co-constitution with the river, and also a reflection of the entanglement of senses, location and material informing this politics (Neimanis 2012, 2013; Singh 2013).

The river regains its waters from the canal and flows onto Talavera de la Reina. Irrigation communities here sprung up after the Spanish Civil War using water from the Alberche River. Yet even this tributary of the Tagus suffers from low flow due to a transfer of its headwaters to Madrid. The river ran dry here in 2006.

The Tagus river runs its course from the region of Castille la Mancha and into Extremadura, where hydroelectric power reigns supreme. The river becomes a series of reservoirs, inundating villages while ancient Roman temples sit atop the view. It receives most of its flow from tributaries. Irrigation associations were created during the Francoist dictatorship, despite low yields since the soil is poor. This was described as a policy of internal colonisation. Even here, the river has its more-than-human witnesses in areas where it flows unimpeded, with hundreds of vultures flying above it in Monfragüe. Yet, the human persists. The river is held in reservoirs throughout this region to ensure that treaty-bound 2,700 hm^3 per year of water are given over the border into Portugal (Escudero Gómez & Martín Trigo, 2020).

But what of the Tagus' waters that flow in the other direction, towards the Levant and the Mediterranean? After the transfer, it mixes with that of the Jucar and Segura Rivers and is spread out along irrigation channels from the Murcia region to the provinces of Alicante and Almeria. When reaching intensely irrigated fields, it is further mixed with groundwater, desalinated water, and treated wastewater, flooded across fields or dripped through plastic pipes, becoming the subject of conflicts over water and farmland (Greenpeace, 2017), and discourse over regional rights to the Tagus. Some of it seeps across the region and into the Mar Menor saltwater lagoon, increasingly afflicted by processes of eutrophication, before entering the Mediterranean. Some also reach the last stretch of the Segura River where it supplies centenary irrigation institutions, which trace their origins to medieval Muslim Spain. For them, their canals hold live and dead water[5] depending on its location within the irrigation network. Irrigation farmers on both ends of the transfer embody divergent narratives over how the waters of the Tagus are best used—whether for intensive and modern forms of irrigated agriculture, or to see its waters as live or dead depending on the canal it flows within. The Tagus then becomes entangled with the European polity as it becomes embodied in the fruits and vegetables that are exported throughout Europe.

Conflicting Meanings and Contradictions of Interdependent Waters

The Tagus exemplifies the meeting of multiple discourses, social and material relationships and embodied subjectivities across many hydrosocial territories formed by water transfers. The river and its tributaries

[5] "Live" water is water that is brought into fields to irrigate them via canals from a source (river, spring, well) while "dead" water is collected from fields after irrigation into canals (Morales Gil et al., 2005). Dead water can be reutilized further downstream (becoming live again) or is brought back into a river.

are a living, historical and contemporary relationship between humans and more-than-humans, the living and the deceased, the material and the discursive. As Puig de la Bellacasa (2017) argues, material practices and ethical dispositions are inseparable, but at the same time they are non-innocent. Multiple conflicting meanings of water, as resource or ecosystem, and the aesthetics and emotions ascribed to it, ranging from pride to sadness to anger, envelop and co-constitute what is a deceivingly simple word, *river*. The politics of interdependency with water, illuminated in these often-contradictory meanings and relations with the river, enables us to reconsider the more-than-human as active beings in what is considered ethical practice, rather than as simply subjects of ethics (Plumwood, 2012).

By highlighting my own relationship and partial understanding of the river, I recognise my own affects and emotions in my personal history, giving due respect to other subjects inhabiting and living with the same river. This requires looking at how capturing emotional communication between human and more-than-humans takes place (González-Hidalgo & Zografos, 2019), and similarly, understanding the politics of subjects' own understanding of how they care for the same shared river and believe that their actions are "life affirming" (Singh 2013, p. 190).

In this respect, the Tagus can be conceived as a tortured example of a conflict between different understandings of Singh's life affirmative actions that spills out and creates multiple waters—urban wastewater, ecosystem water, potential energy water—within a river, and in turn spills onto other more-than-humans—soils, cities, energy systems—that cross multiple scales. Activists denounce the water transfer to Spain's levant as ecologically crippling, while irrigation farmers in the levant as well as in the Tagus basin are upset by being cast as abusers of the river, while they provide food to millions. Water connects the human scale to all other scales of life (Neimanis, 2012). Recounting the complexity of the Tagus's multi-scalar nodes helps record what polluting activities and those in positions of power have done to territories, humans and more-than-humans as well as the unequal impacts of these actions. Recording and narrating illuminates these stories and histories into contemporary politics, thereby giving possibilities for different directions of political action in the constant co-becoming with the river.

With my partial understanding of the Tagus, I draw out the struggles for dominance and memories of different versions of the Tagus. My telling casts a new light on how activists, irrigation farmers, politicians and civil servants fight over articulating different meanings of life-giving and taking on the river, who is the true steward of water and therefore of land and territory, and who is safe-guarding (human and more-than-human) life (Singh, 2013). This problematizes the hydrosocial territory perspective (Boelens et al., 2016); from a partial and situated understanding, I trace part of the webs of relations among identified subjects—myself, my family, farmers, activists—as co-constituting material flows. Yet the webs of relations are never fully knowable.

Similarly, the fight among actors articulating different meanings is also an exclusionary fight over subjective human values that often affords little compromise. Recognising other value systems allows me to judge what and how other human and more-than-human subjects are valued, and requires me to choose with who/what and how I want to belong (Rowe, 2005). It also forces me to identify the consequences of other value systems, their material and discursive articulations, which lead to outcomes that go against my own ethics of more-than-human life. Belonging is invariably nested in divergent and conflicting understandings of territories, the subjects within, and the forces that subjectify. A more-than-human co-becoming sees me/us as interdependent, even if this interdependence is anything but innocent, and the river carries the consequences of this relational web (Neimanis, 2012) as nitrites and ammonia.

I as researcher co-become (Bawaka Country et al., 2013) with the Tagus as I write the interdependent histories and relationships of the multiple human subjects constantly relating *with* the Tagus as more-than-human. That this co-becoming is fraught also shows how "Country" is not innocent; rather, it can support and harm life, reflecting flows of power that are multifaceted and consequential for subjects and ecologies (Nightingale, 2011; Swyngedouw, 1999). My own experience and memory inform my situated politics. Interdependency and co-becoming with the Tagus is relational; it is a political and ethical positioning of the self vis-à-vis the more-than-human river, and against socionatural articulations that subject it and us to exclusionary and oppressive ways of living. The smell of ammonia is a testament to, and result of, the uneven politics that co-constitutes the Tagus, and a call for rethinking the ethical implications of interdependent embodied subjectivities of/with the Tagus.

Ethics of More-Than-Human Interdependencies

These situated narratives of co-becoming with water illustrate the politics of interdependencies and their ethical implications. The ebbs and flows of the Tagus River are implicated in the lives and deaths, the thriving and suffering of people and more-than-humans, all of whom shape the fluid waterscapes of the river both materially and discursively. In Pravah, Sakarbai's water has turned "bad" and the tap water has health implications for human and more-than-human beings who use it. Despite this, there are benefits from the use of wastewater. Flowing with the increased, more reliable water supply are economic and social benefits for many in the farming communities. The multiplicity of different waters in both cases are therefore infused with ethics through their situated more-than-human interdependency. Rather than being considered moral or immoral, these relationships with multiple waters are understood and experienced as part of wider webs of ethical significance in everyday life (Puig de la Bellacasa, 2017). The co-mingling of pasts, presents and potential futures in such waterscapes is therefore also knotted with ethical relations that inform ways of living with the multiplicity of water as it moves, laden with power, through lands, bodies and infrastructures (Rose, 2012). These illustrations of more-than-human interdependencies demonstrate that rather than occurring in an "encounter", interdependencies are always ethical because they emerge through relational assemblages; entangled lives carry responsibility to others, giving both life and death.

When ethics are understood in this way, as enmeshed in multi-species everyday practices, embodied engagement within the world, and more-than-human relational knowledge formation (Richardson-Ngwenya & Nightingale, 2018), this legitimises and makes necessary an acknowledgement of more-than-human interdependencies as crucial to the maintenance of life (Puig de la Bellacasa, 2017). Drawing on the work of Levinas and his student Hatley, feminist philosopher Deborah Bird Rose (2012) exemplifies the workings of more-than-human ethical relations by extending the idea of ecological relationality beyond spatially configured power dynamics and locating ethics in time, a move which helps deepen the ethical and justice dimensions of the two water narratives here. Mobilising the work of ecologists on "flying foxes and their co-evolved myrtaceous mutualists", Rose (2012, p.135) argues that the webs of mutualism that occur between the myrtaceous trees and flowers which attract flying foxes, who in turn assist in the pollination of various woodland and rainforest species, can be understood as ethical in that they maintain each other, themselves and their future "selves" or generations (see also Zwarteveen & Boelens, 2014 and their conceptualisation of 'socio-ecological justice'). Ethics becomes the ontological condition of the ebb and flow of life and death; it is a responsibility, or maintenance of life, between different beings.

These insights bring us to questions of care and responsibilities once we have narrated co-becomings such as those above. What Rose (2012, p.136) terms "multispecies knots of ethical time" are the intertwined histories, presents and futures that are embodied in the practices of species as they move through the world. This ethical intertwining of bodies and ecologies over time and space relates to the feminist stance on the ethics of care as "a vital interweaving web of life" (Puig de la Bellacasa, 2017, p. 4). Following Tronto (2015), Puig de la Bellacasa (2017, p. 6), maintains a distinction between ethical dispositions and material practices while calling for them to be considered inseparable. In doing so she articulates more-than-human interdependency as an ethics of care:

> [...]'ethics' in an ethics of care cannot be about a realm of normative moral obligations but rather about thick, impure, involvement in a world where the question of how to care needs to be posed. That is, it makes ethics a hands-on, ongoing process of recreation of 'as well as possible' relations and therefore one that requires speculative opening about what a possible involves [...] unthinkable as something abstracted from its situatedness.

Thinking of ethics as embodied and situated, helps to decentre the human, and therefore moral discourse, in more-than-human interdependencies. As a speculative politics, this ethics of care is "non-innocent", meaning it is both troubled by its situatedness and co-constituted through complex more-than-human intra-action (Barad, 2007). In Pravah this manifests in the way the wastewaterscape and its social, economic and embodied consequences cause harm as well as material benefits, a contradiction that plays out as women farmers navigate the different scales of the wastewaterscape in everyday life. These ambiguities are further exemplified in the paper by Bawaka Country and colleagues (2013), where they reconceptualise natural resource management through the Aboriginal indigenous ontology of co-becoming. *Wetj*, a practice of reciprocal sharing, care and "intra-action" (Barad, 2007) between all beings,

"...springs from and supports a Yolŋu ontology of co-becoming which sees all beings, including human beings, as coming into existence through relationships" (Bawaka Country et al., 2013, p. 187). Or in other words, beings "only exist, be-come vibrant, powerful and important, through relationships" (ibid., 2013, p. 189). They demonstrate how the perspective and meaning of natural resource management changes when agency and communicative status is given to Country. Rather than assuming humans as the primary care-givers and care-takers of Country, the maintainers of morality, Country is seen as a sentient, multifaceted assemblage of beings that sometimes cares for humans and non-humans, sometimes harms, but is always interdependent, in a mutual, relational co-becoming. As narrated above, the entangled waterscapes of the Tagus demonstrate the ways that water as an ambiguous care-giver and care-taker of multiple more-than-human assemblages, also engender a non-innocent process of co-becoming. These often-contradictory relations of mutuality also emerged clearly in the first section, as we unfolded how waters and gendered subjectivities co-constitute one another—or intra-act—in the wastewaterscape of Pravah in ambiguous, non-innocent ways.

If ethics are embedded in the everyday life of all beings, which unfolds over and through diverse temporalities (Rose, 2012) and socionatural flows of power (Nightingale, 2011), the multiplicity and mobilities of water as it percolates through rocks, soils and bodies, is itself ethically infused. The complexity of ethics within this wider frame of "multispecies knots" cannot therefore be reduced to analysis of a particular encounter between "human" faces, nor as Barbara Davy (2007) argues, the requirement of human recognition or legitimisation of the other that this moment implies. The waters of Pravah and the Tagus, in their multiple forms, both continuously enable and disable life; they both take and give life (Bawaka Country et al., 2013). Therefore, to build on Rose's (2012) and Davy's (2007) arguments, ethics, we suggest, are provoked, but not through a particular encounter between knowing ethical selves. Rather ethical relations emerge through dwelling *within* more-than-human interdependencies.[6] Thinking with the waters of Pravah and the Tagus demonstrates the uneven politics of dwelling within more-than-human interdependencies. Teasing out the ways in which water entangles humans

[6] For further discussion on the more-than-human politics of dwelling see Tim Ingold (2005).

enables us to think beyond human-centric ethical practices and implications, and to conceive of ethics as embedded within the co-constitution of embodied subjectivities, within processes of more-than-human co-becoming, creating for us new questions about interdependencies, caring and the kinds of narratives we tell about those co-becomings.

Conclusions

In this chapter we have explored different ways of grappling with the more-than-human, focusing on the politics of interdependencies through situated entanglements with water. We have done so by juxtaposing two cases that are geographically distant and working on different scales of operation. Human-water relations in the village of Pravah, in Maharashtra, and the flow of the Tagus River, in Spain, shape differentiated and ever-changing embodied subjectivities (including those of the researchers): a process of co-becoming that has specific implications for how we understand and engage with more-than-human interdependencies and therefore ethics.

Drawing on FPE and other allied disciplines, we started our journey of grappling with the more-than-human by acknowledging the importance of unpacking nature/society binaries, the symbolic and material boundaries that maintain them and the power dynamics in which they are enrolled. We thus framed more-than-human interdependencies from an understanding of relationality as entangled, multiple and situated (Nightingale, 2011; Rowe, 2005). Understood as interdependent and entangled, the more-than-human forces us to look beyond distinct individuals, supposedly separate from ecologies and other beings. Therefore, it becomes important to attend to the specific and situated characteristics of such relations as the basis for theorising. To unpack these characteristics of more-than-human interdependency, we found it useful to interweave FPE with STS and post-human scholarship: it helped us attend to the specific affordances of water (water percolates, contaminates, transports sediments and algae, etc.), which play a role in shaping differentiated gendered embodied subjectivities and, conversely, how water is signified.

Two main points emerged as we walked through Pravah with women farmers and along the flows of the Tagus River. Firstly, more-than-human relations, and especially human-water relations, are ambiguous, troubling and never innocent. They create differentiated possibilities of well-being and illness, both for human and more-than-human actors.

Secondly, as the authors, we actively co-enact the more-than-human relations we describe and theorise about. And this has ethical and political implications, not least because what we can say about the more-than-human is always partial, mediated by the social, by our (human) ways of making sense of the world. For instance, the focus on the interdependency of women farmers and (waste)water in Pravah is a specific political choice to challenge the processes through which experiences of women farmers from multiple castes, their knowledges and practices in relation to (waste)water, are marginalised and/or silenced; and to interweave feminist struggles with environmental concerns in Maharashtra. The story of the Tagus takes on a different positionality, embodied within the hydrosocial territory it narrates. The process of knowing and becoming with the Tagus invariably collides with other divergent forms of co-becoming that create an emotional and sensorial rejection of processes that make specific territories. Nonetheless this same process also informs a situated politics, one which demands an acknowledgement of how everyone and everything that is part of the more-than-human co-become in ways that defy anthropocentric modes of (dis)ordering the world (Halberstam, 2020).

This chapter therefore illuminates the ways water relations are ethically infused processes of co-becoming. In illustrating the emergent ambiguity of more-than-human interdependency, the chapter contributes to troubling the neat moral boundaries that constrain the multiplicity of more-than-human ethical relations within spheres of privileged human moral reasoning (Richardson-Ngwenya & Nightingale, 2018; Tronto, 1993). Through highlighting the politics of interdependencies that animate the multiple waters along the Tagus and in Pravah, and exploring our own role as researchers in the narration of such, we have also sought to question the nature of ethics as something bounded by human reason. These particular narrations of co-becoming with water highlight the importance of empirically unpacking the complexity of embodied waters and how they shape not only economic and agricultural possibilities, but also every day, embodied senses of well-being, illness and change in contemporary waterscapes. Exploring more-than-human relations therefore involves thinking with the situatedness and embodiment of such relations, thus revealing the ways in which power flows and plays across what often appear as distinct bodies, spheres or scales. In doing so, these intersecting relations and an understanding of our own entanglement in them as researchers, contributes to FPE, STS and philosophies that seek to disorder or blur the boundaries that constrain life and deeper understandings of its more-than-human interdependencies.

Funding: This chapter was funded by the Wellbeing Ecology Gender and cOmmunities Innovation Training Network (WEGO-ITN) funded by the European Union's Horizon 2020 research and innovation programme under the Marie Sklodowska-Curie grant agreement No. 764908-WEGO 2018-2021.

References

Abrahamsson, S., Bertoni, F., Mol, A., & Martín, R. I. (2015). Living with omega-3: New materialism and enduring concerns. *Environment and Planning D: Society and Space, 33*(1), 4–19.

Agarwal, B. (2003). Gender and land rights revisited: Exploring new prospects via the state, family and market. *Journal of Agrarian Change, 3*(1–2), 184–224.

Ahmed, S., & Zwarteveen, M. (2012). Gender and water in South Asia: Revisiting perspectives, policies and practice. In M. Zwarteveen, S. Ahmed & S. R. Gautam (Eds.), *Diverting the flow: Gender equity and water in South Asia* (pp. 3–30). Zubaan.

Barad, K. (2007). *Meeting the universe halfway*. Duke University Press.

Baviskar, A. (2007). *Waterscapes: The cultural politics of a natural resource*. Permanent Black.

Bawaka Country et al. (2013). Caring as country: Towards an ontology of co-becoming in natural resource management. *Asia Pacific Viewpoint, 54*(2), 185–197.

Bhat, S. (2016). Deserted and widowed women's struggles for land and livelihood: A case from Maharashtra. In B. Fernandez, M. Gopal & O. Ruthven (Eds.), *Land, labour and livelihoods* (pp.73–88). Palgrave Macmillan.

Boelens, R., Hoogesteger, J., Swyngedouw, E., Vos, J., & Wester, P. (2016). Hydrosocial territories: A political ecology perspective. *Water International, 41*(1), 1–14.

Bossenbroek, L., & Zwarteveen, M. (2018). New spaces for water justice? Groundwater extraction and changing gendered subjectivities in Morocco's Saïss region. In R. Boelens, T. Perrault & J. Vos (Eds.), *Water justice* (pp. 330–345). Cambridge University Press.

Butler, J. (1990). *Gender trouble: Feminism and the subversion of identity*. Routledge.

Davy, B. J. (2007). An other face of ethics in Levinas. *Ethics and the Environment, 12*(1), 39–66.

Elmhirst, R., Siscawati, M., Basnett, B. S., & Ekowati, D. (2017). Gender and generation in engagements with oil palm in East Kalimantan, Indonesia: Insights from feminist political ecology. *The Journal of Peasant Studies, 44*(6), 1135–1157.

Escudero Gómez, L. A., & Martín Trigo, A. (2020). La Gestión Pública de Una Obra Hidráulica Compleja, Entre El Marco Internacional y Los Intereses Regionales Privados: El Caso Del Trasvase Tajo-Segura (España). *Relaciones Internacionales, 45*, 327–344.

Fenske, M., & Norkunas, M. (2017). Experiencing the more-than-human world. *Narrative Culture, 4*(2), 105–110.

Gnana, S. R. (2018). Caste system, Dalitization and its implications in contemporary India. *International Journal of Sociology and Anthropology, 10*(7), 65–71.

González-Hidalgo, M., & Zografos, C. (2019). Emotions, power, and environmental conflict: Expanding the 'emotional turn' in political ecology. *Progress in Human Geography, 44*(2), 235–255.

Greenpeace. (2017, August 24). *La Trama Del Agua En La Cuenca Del Segura Diez Años Después*. Retrieved 4 July 2022 from http://archivo-es.greenpeace.org/espana/es/Informes-2017/Agosto/La-trama-del-agua-en-la-cuenca-del-Segura/

Halberstam, J. (2020). *Wild things: The disorder of desire*. Duke University Press.

Haraway, D. J. (2003). *The companion species manifesto: Dogs, people, and significant otherness*. Prickly Paradigm Press.

Harris, L. M. (2006). Irrigation, gender, and social geographies of the changing waterscapes of southeastern Anatolia. *Environment and Planning D: Society and Space, 24*(2), 187–213.

Hernández-Mora, N., Del Moral Ituarte, L., La-Roca, F., La Calle, A., & Schmidt, G. (2014). Interbasin water transfers in Spain: Interregional conflicts and governance responses. In G. Schneier-Madanes (Ed.), *Globalized water: A question of governance* (pp. 175–194). Springer.

Ingold, T. (2005). Epilogue: Towards a politics of dwelling. *Conservation and Society, 3*(2), 501–508.

Isaacs, J. R., & Otruba, A. (2019). Guest Introduction: More-than-human contact zones. *Environment and Planning E: Nature and Space, 2*(4), 697–711.

Jagtap, S. S., & Manivanan, R. (2019). Water pollution status of Mula-Mutha rivers in Pune city: Review. *International Journal of Trend in Scientific Research and Development (IJTSRD), 4*(1).

Karpouzoglou, T., & Zimmer, A. (2016). Ways of knowing the wastewaterscape: Urban political ecology and the politics of wastewater in Delhi, India. *Habitat International, 54*, 150–160.

Krishna, S., & Kulkarni, S. (2019). Gender and water: Why we need alternatives to alternative discourses. In K. J. Joy & S. S. Janakarajan (Eds.), *India's water futures: Emergent ideas and pathways* (pp. 235–252). Routledge.

Leonardelli, I., Kemerink-Seyoum, J., & Zwarteveen, M. (2022). Obliqueness as a feminist mode of analysing waterscapes: Learning to think with overflows. *Environment and Planning E: Nature and Space.*

Leonardelli, I., & Tozzi, A. (forthcoming). From scattered rainwaters to contaminated wastewaters: A feminist study into how 'water worlds' shape processes of rural agrarian transformations in Maharashtra, India. In M. Zwarteveen et al. (Eds.), *Gender and water governance.* Routledge.

Mehta, L. (2014). Water and human development. *World Development, 59*, 59–69.

Mehta, L., & Karpouzoglou, T. (2015). Limits of policy and planning in peri-urban waterscapes: The case of Ghaziabad, Delhi, India. *Habitat International, 48*, 159–168.

Morales Gil, A., Rico Amorós, A. M., & Hernández Hernández, M. (2005). El Trasvase Tajo-Segura. *Observatorio Medioambiental, 8*, 73–110.

Moreno, L. U. (1980). La Fresa en Aranjuez. *Estudios Geográficos, 41*(159), 211.

Morote, A. F., Hernández, M., Rico, A. M., & Eslamian, S. (2020). Interbasin water transfer conflicts. The case of the Tagus-Segura Aqueduct (Spain). *International Journal of Hydrology Science and Technology, 10*(4), 364–391.

Neimanis, A. (2012). Hydrofeminism: Or, on becoming a body of water. In H. Gunkel, C. Nigianni, & F. Söderbäck (Eds.), *Undutiful daughters: New directions in feminist thought and practice* (pp. 96–115). Palgrave Macmillan.

Neimanis, A. (2013). Feminist subjectivity, watered. *Feminist Review, 103*(1), 23–41. https://doi.org/10.1057/fr.2012.25

Neimanis, A. (2017). *Bodies of water: Posthuman feminist phenomenology.* Bloomsbury Publishing.

Nelson, I. L. (2017). Interspecies care and aging in a gorilla 2.0 world. *Geoforum, 79*, 144–152.

Nightingale, A. J. (2006). The nature of gender: Work, gender, and environment. *Environment and Planning D: Society and Space, 24*(2), 165–185.

Nightingale, A. J. (2011). Bounding difference: Intersectionality and the material production of gender, caste, class and environment in Nepal. *Geoforum, 42*(2), 153–162.

Nightingale, A. J. (2013). Fishing for nature: The politics of subjectivity and emotion in Scottish inshore fisheries management. *Environment and Planning A, 45*(10), 2362–2378.

Nyantakyi-Frimpong, H. (2019). Visualizing politics: A feminist political ecology and participatory GIS approachto understanding smallholder farming, climate change vulnerability, and seed bank failures in Northern Ghana. *Geoforum, 105*, 109–121.

Padhi, R. (2012). *Those who did not die: Impact of the Agrarian crisis on women in Punjab.* SAGE Publications.

Pickering, A. (2009). New ontologies. In A. Pickering & K. G. Guzik (Eds.), *The mangle in practice science, society, and becoming* (pp. 1–16). Duke University Press.

Plumwood, V. (2012). *The eye of the crocodile*. Australian National University E Press.

Puig de La Bellacasa, M. (2017). *Matters of care: Speculative ethics in more than human worlds*. University of Minnesota Press.

Richardson-Ngwenya, P., & Nightingale, A. J. (2018). Diverse ethics for diverse economies. In C. Bauhardt & W. Harcourt (Eds.), *Feminist political ecology and the economics of care: In search of economic alternatives* (pp. 131–161). Taylor & Francis.

Rose, D. B. (2012). Multispecies knots of ethical time. *Environmental Philosophy, 9*(1), 127–140.

Rowe, A. C. (2005). Be longing: Toward a Feminist Politics Of Relation. *NWSA Journal, 17*(2), 15–46.

Singh, N. M. (2013). The affective labor of growing forests and the becoming of environmental subjects: Rethinking environmentality in Odisha, India. *Geoforum, 47*, 189–98.

Still, E. (2022). Untangling agricultural ethics: Women's farming collectives as alterbiopolitics. *ASIEN, 160*(161), 17–42.

Sultana, F. (2009). Fluid lives: Subjectivities, gender and water in rural Bangladesh. *Gender, Place and Culture, 16*(4), 427–444.

Sultana, F. (2011). Suffering for water, suffering from water: Emotional geographies of resource access, control and conflict. *Geoforum, 42*(2), 163–172.

Swyngedouw, E. (1999). Modernity and hybridity: Nature, regeneracionismo, and the production of the Spanish waterscape, 1890–1930. *Annals of the association of American Geographers, 89*(3), 443–465.

The Multispecies Editing Collective. (2017). Troubling species: Care and belonging in a relational world. *RCC Perspectives: Transformations in Environment and Society, 1.*

Tronto, J. C. (1993). *Moral boundaries: A political argument for an ethic of care*. Routledge.

Tronto, J. C. (2015). *Who cares?: How to reshape a democratic politics*. Cornell University Press.

Tsing, A. (2015). *The mushroom at the end of the world: On the possibility of life in capitalist ruins*. Princeton University Press.

Zwarteveen, M., & Boelens, R. (2014). Defining, researching and struggling for water justice: Some conceptual building blocks for research and action. *Water International, 39*(2), 143–58.

CHAPTER 7

Meanings and Practices of Care in Feminist Political Ecology: An Intergenerational Conversation with Khayaat Fakier and Wendy Harcourt

Marlene Gómez, Anna Katharina Voss, and Eoin Farrelly

Introduction

This wide-ranging conversation circles around meanings of care from different academic disciplines as well as feminist activist practices in order to sketch out some of the diverse understandings of care in feminist

M. Gómez (✉) · A. K. Voss · E. Farrelly
Freie Universität Berlin, Berlin, Germany
e-mail: marlene.gomez@fu-berlin.de

E. Farrelly
Department of Sociology & Human Geography, University of Oslo, Oslo, Norway

A. K. Voss
International Institute of Social Studies, The Netherladns & Fondazione Pangea Onlus, Italy, The Hague, The Netherlands

W. Harcourt et al. (eds.), *Contours of Feminist Political Ecology*, Gender, Development and Social Change,
https://doi.org/10.1007/978-3-031-20928-4_7

political ecology (FPE). The intergenerational conversation is based on questions posed by Marlene, Anna and Eoin, all early career scholars, to Khayaat and Wendy, two teachers and writers who have long been engaged in FPE, one in the Global South and the other in the Global North. Both Khayaat and Wendy have been involved in feminist activism as well as academic institutions and come from different academic entry points to care. The conversation touches on how care has different meanings according to disciplines and experience. It looks at meanings of care from diverse scholarly positions—from feminist economics, sociology, critical development studies, queer ecology and decolonial approaches—in order to consider how FPE research and teaching can be done *with care*. The responses of the two scholars to the questions posed suggest the ways in which different meanings of care across generations filter into FPE from both a personal and political perspective—including Khayaat and Wendy's unique way of practising a feminist ethics of care in the classroom. While the conversation is only able to partially cover the importance of care to FPE, it sets out some of the basic questions that FPE scholars continue to ask about the politics and ethics of care towards the human and the more-than-human. We have tried to signpost the different threads in the conversation through sub-titles to help the reader follow how the conversation moved.

Positioning the Conversation

Marlene: As feminist political ecology (FPE) scholars, how do you relate to the different meanings and practices of care?

Khayaat: I relate to care at different levels, from my position as a scholar but also as an engaged human being in society and social movements. As an FPE scholar, I relate to care as a concept developed in feminist literature which raises the questions: How do we value care? Who conducts care? Who should be responsible for care? How do we look at care in public and private institutions? How does care play out in communal activities and as a public good? FPE values care as a fundamental need of humans and societies since we cannot flourish without care. Thus, care encompasses diverse activities including the values that support our mental and political health for social development and the responsibility we have to care for human and more-than-human others so that we may stop thinking of the planet merely as a container of resources.

We need to look at what we consume and how we interact with each other and understand care as an ethical practice.

The big question is: what is sufficient for us to live? People experience wide disparities in access to resources and in their experience of sufficiency and replenishment. The relations we have with resources need to be transformed and rebuilt from an ethical practice of care, where care is understood as a fundamental need, value, and activity. I relate to care not only as a concept to study but also as a practice. A care-full practice means that we need to pay close attention to the needs of others and the impact of our actions in our everyday life as well as in our research methodologies. Another important aspect is to avoid extractivism of epistemologies and ontologies. Doing research from an ethics of care involves thinking of research as a process of collaboration with the research participants as well as one's interaction with peers and negotiation with people in power, inside and outside the academy.

Wendy: As a feminist political ecologist I relate to care, like Khayaat, in various ways. Here I would like to highlight five. First, I relate to care as a principle of how I live my life, thinking through how I care for and with others, retrieving Tronto's (1993) debate on "caring with others". This is still an ongoing process through which I am learning individually and with others. I seek to be always aware of my positionality, specifically my race, age and educated class privileges.

Second, we need to discuss the politics of care, which determine who is able to care for whom. In this regard, I edited a book with Christine Bauhardt that built both on feminist political ecology and feminist political economy, examining the politics of care from these two similar but different viewpoints. In the book we look at how the concept of care is not just related to social reproduction but also to caring with more-than-human others and for the planet. There were several interesting tensions around how feminist economists understand care (how to measure the importance of care work) in contrast to feminist ecologists (how to understand relations of care with humans and more than human others)—this is an important debate within feminist theory and practice.

A third important area where I have learnt about care is in conversations around degrowth, where care is seen as a value that can displace economic growth and greed. Engaging with degrowth reflects my lifelong personal, political, and academic struggle with economists and traditional views on economics, since care as a value requires a very different understanding of the economy, labour and intersectional determinants—care is relational and complex.

Fourth, I see care as needing to embrace the more-than-human. Personally, that is difficult as I live in an urban context, embedded in Western-centric, Eurocentric, and human-centric dynamics. So, while I am fascinated by the theory of the more-than-human and how we are entangled in multispecies interactions, in practice I need to be more hands on how to care for more-than-human others. I love to be in nature, but at the same time, I feel at a loss when faced with the harshness of what is happening to our environment with climate change and ecological destruction. When we talk about care for the more-than-human, I know we need to go beyond the image of nature somewhere out there and admit how enmeshed we are in our environment. But in the end what I feel is an unease at caring with and for nature. What does it really mean, personally and collectively, to resist climate change and other socio-natural disasters and crises, particularly as someone living in urban environments? What kind of socionatures do I inhabit and shape?

Finally, coming from Australia as a white settler feminist, I am trying to understand the meaning of caring for Country (a term from Indigenous Australia). Living in Italy and working in The Netherlands since my late 20s, I have lived far away from Australia, but I am shaped by that history. Every visit, I see changes to non-Indigenous Australian politics as more people (mostly on the left) become aware of what Indigenous knowledges can teach them about their environment and the damage white settler life has done. I feel there is much more learning needed if there is to be a meaningful dialogue across race, class and gender around Country. As an ally to Indigenous Australians, I want to learn more about care with Country, even if I am mostly only able to engage virtually and via translation.

Different Understandings of Care in Academic Debates

Marlene: What are the different understandings of care in feminist and environmental justice literature and what are the contradictions?

Khayaat: It is important to understand how sociologists, feminist economists, development scholars, and others look at care, including in terms of the differences among literatures. The difference sometimes lies in how and where we place the emphasis when analysing a problem. If one looks at the sociological literature, the attempt to understand care mostly comes from questioning who and what needs care and who and what conducts care, a perspective found mostly in a sociohistorical and economic approach.

Feminists working on care, such as Jody Heymann (2007), Bridget Anderson (2021), and others look at care as a human concern through the study of maternity. Their work disrupts the idea of a "universal need for care" through looking at class differences and how they shape global care chains. Through looking at care from a social class perspective, feminist economists and development scholars, such as Nancy Folbre (1995, 2014) and Shahra Razavi (2011a), look at the institutions of care; that is who cares and who benefits from care. Other scholars look at care from an interdisciplinary perspective, looking mostly at how care relations are found among people and the environment in everyday life. Wendy's work comes to mind (2009, 2017), and also Jacklyn Cock's (2021) and Nora Rathzel's (Räthzel & Uzzell, 2019), scholars I have worked with that develop an approach to care for the environment. Although these scholars work on the topic of care, there are many contradictions and tensions in their literature that are visible in how they include intersectional issues of race, class, gender, sexual orientation as well as the roles that the environment and social justice movements play. In my own work, I try to be more eclectic and therefore more pragmatic. If we're talking about, for instance, the survival of humans and more-than-humans, we need an interdisciplinary approach. We need to work with the contradictions and broaden the dialogue across academic literatures as well as listen to ways of understanding the world outside the canon of academia. We identify these tensions in those who focus on state-centric approaches versus those who take a bottom-up approach and are critical of the state, instead looking to social movements. It is not that one is right and the other wrong; they both have something important to say about the politics of care. We need to bring those perspectives together and work through the contradictions.

Wendy: I am thinking on similar lines to Khayaat about the need to see what different disciplines offer, and how FPE offers a transdisciplinary approach to care. Feminists, whether working in sociology, economics or development studies, examine who is doing the care, for whom and where. What is important is that their studies make care visible and underline how care is crucial to keep our lifeworld going.

Economists like Folbre (1994) have asked who cares for the kids and shows how economies need to recognise that reproductive work underscores all productive activity. Razavi's work in gender and development has produced her model of the care diamond (2007) which conceptualises how the provision of care includes the family/household, markets, the public sector and the not-for-profit sector (including voluntary and community provision). Both argue that care (whether paid or unpaid) is crucial to human well-being and to social and economic development, as part of the fabric of society. Feminist economists underline it is important not only to count and measure care but to change economic thinking to revalue care work as crucial to the economy. In another important theoretical move, they also invite us to go beyond household chores to see love and emotions as crucial to everyday life and survival.

Feminist ecologists, on the other hand, speak about care for nature and environmental justice. They critique environmentalists' focus on conserving biodiversity as more important than peoples' habitats. As ecofeminists have stated, the question is not are we too many but why are we too greedy? The relationship between population and environment is a complex and passionate debate among ecologists and economists. Haraway's rallying cry "make kin not babies" proposes that we consider the population question in a different way—not about numbers and scarcity of resources but about the importance of care for more-than-human relations (see the chapter on population in this book). Studies on gender relations and biodiversity by Joni Seager (1993) for example also show the importance of gender, environment and gender relations. She points to the gendered division of knowledge and labour demonstrated through women in different cultures knowing and caring for particular plants and animals. FPE scholars Christa Wichterich and Giovanna Di Chiro have also looked at care in relation to global care chains, environmental concerns around the green economy, and in relation to degrowth (see Chapter 8 on degrowth in this book).

One way to do care-full research is to open up academia to different forms of knowledge through storytelling. Telling stories about how we live with more-than-human others builds on oral traditions. Powerful origin stories remove the dominant western narrative of rational "economic" man and ask us to listen to the more-than-human, paying attention to how humanity is embedded in nature, entangled with the land, water, forests. Post-development scholars like Arturo Escobar (2020) speak of the pluriverse (rather than universe) to recognise the many ways of knowing, and thinking-feeling or *sentipensar*, as key to knowledge production. Telling stories about the history of place, how place evolves and how we feel and sense those changes, help us to envisage a pluriversal rather than a single, hegemonic story of progress, modernity and development. The historical and ongoing stories of violence in colonial and post-colonial societies in white settler countries and the Global South illustrate how colonial rule and global neoliberal capitalism have not only led to human suffering and exploitation but also planetary suffering. We have created life-worlds that fail to care for people or for the planet: climate crisis, flooding, fires, that make life unliveable. Indeed, sometimes it seems like a form of revenge by the Earth. That might sound extreme and emotional, but I think we do need to add to rational, scientific, and measured analysis the deeper reasons for what is causing our collective fear in order to learn how to care in a more holistic way. If we stay with our emotions and fears and begin to tell other, more meaningful stories about care we might learn to position care at the centre of our lives and move away from the violence of extractivist Eurocentric technoscience.

Learning About Care in Different Social Contexts

Anna: Thank you both for taking our readers on such a nuanced journey through the complexity of understandings of care, and in pointing out how FPE and kindred social sciences are asking uncomfortable questions and unpacking our individual and societal conditioning. Challenging our anthropocentric worldviews can be unsettling and requires humility. But the curiosity to (re)learn other ways of inhabiting the planet also opens up fascinating paths to a more embodied awareness of our more-than-human interdependence. The slogan "We are not defending nature, we are nature defending itself", endorsed by Indigenous environmental activists comes to my mind here.

Khayaat: When I was thinking about this discussion, I was thinking about embodiment and a very visceral experience of care when we become disconnected from our relation to the more-than-human. I agree with Wendy, we are experiencing a very violent embodiment of a disconnection among the human and more-than-human. If one looks at nutrition levels in South Africa, there is violence due to malnutrition, with people slowly starving. That has a very real embodied impact. The latest research, influenced by COVID-19, is that nearly 30% of children under the age of five in South Africa are malnourished. Not caring for the environment has a physical impact on bodies, as well as mental, social and emotional development. In South Africa a disregard for the impact of extractivism (such as mining and industrial agriculture) on the soil, climate and ecology of food production breaks the caring connections between humans and non-human-others. Not only does this affect the diversity and availability of food and result in under-or malnutrition, but care for the environment is also displaced, and humans become alienated from their more-than-human others.

This is quite significant in South Africa, which is seen as the most developed country on the African continent. South Africa used to be the food basket of the world, but now it holds out begging bowls. One of the important reasons is the influence of industrial agriculture on the country's ability to feed its own people. We need to talk about a care crisis in relation to industrial agriculture, civil war, the displacement of people and the development of crops which are grown for export rather than feeding people. In our economic growth-driven world, commodities are more important than the sustenance of people. A lack of care by governments leads to a lack of focus on inequality and equal access to nutrition, and to the poor physical health and growth of children, as well as rise of obesity, which goes hand-in-hand with under-nutrition and malnutrition. We need to understand how industrial development processes lead to industrial enclosures, as well as what Wendy mentions: how conservation practices forcibly remove people from their land and stop them from growing their own food or producing food for their country.

These processes of modern enclosure, led by massive industrial forms of production, have also led to an urban bias in policymaking. The creation of a modernised, urban, city dweller,—the industrial labourer or a professionalised worker—is reflected in a rural-urban imbalance. Through the modern development process, citizens are increasingly divorced from an embodied sensory engagement with nature. The modern citizen is removed and remote from nature, an individual whose aspirations, work and sensory experience no longer relate to the more-than-human. We need to regain those relations if we are to find sufficiency.

Marlene: I agree with Wendy that storytelling is important. Her questions around which stories we want to hear, what we want to tell, and who is able to tell these stories are crucial. Storytelling is at the core of a feminist practice, and we need to consider this when thinking about our responses to government policies and responsibilities. We have to be mindful that practices of care take place beyond households and are found in our everyday life and at other scales. The question is, how can we translate practices of care into public policies? What are the core points we should need to tackle if we want to ensure policies of care at local, regional, worldwide levels?

Khayaat: When we talk about social policy, we need to think about different aspects: industrial policy, labour policy and environmental policy. Care in South Africa depends on women's volunteer work: women perform vital work in caring for the environment, in caring for orphans, for the disabled and for others. Even if paid, they receive an amount which they can hardly survive on, nor can they afford to train themselves to adequately care for the ill, for example, people living with HIV/AIDS, never mind finding out how to practise self-care.

Policy in the Global South looks very different from social policy in the Global North. Reflecting on what Wendy said about Razavi's care diamond, feminist policy must be about recognising the value that care has for society as a public good, and as a commons. Razavi (2011b) makes a distinction between what happens in the Global South compared to the Global North. Women in the Global North, especially in social democratic welfare states, have been able to develop careers in care, be paid and become emancipated both socially and economically, because the work that they were doing was recognised. This is very different from having care conducted by (women) volunteers. Care work is seen as a side line, even a form of charity. What I would argue is that care should be at the centre of industrial policy and labour policy. When we make social policy about industrial relations and labour conditions, we need to include an understanding of how care for human others and for the environment fit into such policies.

In the Global South, volunteers who are in public works programs are rarely being paid a liveable wage for the very valuable work that they do. What kind of policy could evolve that takes into account better pay and conditions for them? UN Women and the Commission on the Status of Women do take up these concerns in the multilateral space, but it is not a space which is easily accessible to most Global South women, activists or social movements. The value and recognition of care need to emerge from national contexts, where feminist movements put pressure on states by bringing building policy change from the bottom up.

Politicising Our Understandings of Care

Eoin: If care is conceptualised at the global level or scale, what political challenges do you anticipate? How do you suggest they would be navigated and overcome? As we conduct this interview, the international community is preoccupied with the Russian invasion of Ukraine—this is just one example of changing priorities shifting the international agenda away from other global concerns. How could this impact interventions that move towards environmental justice?

Wendy: It is probably rather early to reflect on the impact of the Russian war in Ukraine on global politics. But in terms of the everyday responses to the war in Europe, like in the first wave of COVID, people want to do something to show they care. Many European cities have seen demonstrations, collections of clothes and food and money. Many individuals have invited refugees into their home. Even racialized students who did not get the same official support felt students acted in solidarity with them. I think there is a strong sense of caring and solidarity expressed. It comes down to the interesting question of how transnational you can be and whether acting in place is the way to show real care.

On the other hand, at least for those who have access to media and social media in the West, we see how our fears are individualised as people speak of the rise in mental illness. Even the energy crisis (due to war or climate change) is put in terms of the energy bills individuals will pay and the impacts of inflation on European lives. Caring about keeping European consumer rights without a sense of what larger environmental justice concerns are determining the price rises is not the care we are talking about. We need to shift from conceptualising just ourselves to care with responsibility and with others and feeling that, collectively, we have the agency to do something about it. This type of mental shift will be important if care is really to become central to global approaches to the economy, politics, health and the environment.

Anna: I agree, the question of how to make this shift of values happen is deeply urgent. And as Wendy just mentioned, we need to take into account the role of the media. What makes it to the headlines—whether it is war, or social and economic violence in general, the ongoing environmental collapse or the impact of the pandemic—reflects the absence of care for our planetary well-being. I often feel the tension between falling into despair about the present and what-is-yet-to-come, and at the same time keeping hope that changes are happening at institutional levels and in small, everyday actions of care and solidarity.

Khayaat: I think Wendy is correct in saying that it is a bit early to reflect on what is happening now in Ukraine because of the invasion of Russia. But it is interesting to reflect on what happened during COVID; a situation which was so new to all of us. It cut us off from our normal forms of organising, protesting and engaging with the world because of lockdown restrictions. Especially in South Africa, we couldn't even go outside to exercise, you were limited to your house, in whatever form that took, for several months. Those circumstances amplified the need for care. People wanted to provide care generously even at the cost of their own health. Some women's movements of community-based carers continued to look after other people in their communities, despite the restrictions on their movements. They battled to be defined as essential workers. Even if they weren't paid, they didn't forget the struggle to be recognised and to be valued. They still protested and marched, despite the regulations that people couldn't gather and people couldn't be outside in public.

What I learnt from them was how important it was that they stuck to their values, remembering for whom and why they are organising. That is a very small example, and I certainly don't want to extrapolate it to other contexts, but I think it has a strong message about not losing sight of the value of what movements are doing.

Queer Ecology and Care Otherwise...

Marlene: Talking about the way we engage with the world and also the diverse imaginaries and knowledges we have, I want to raise the topic of queer ecologies. We know that both academic and activist spaces develop diverse proposals to mitigate and tackle the ecological crisis, but these proposals often are heteronormative or they overlook sexual or gender diversities. In this regard, and thinking from a queer ecology perspective, what role should sexual gender diversities play in building a practice of care to contest the ecological crisis we are currently living in?

Wendy: Queer ecology challenges heteronormativity on different levels, it is not just about heterosexual expression and marriage as the norm, it is also about diverse bodies and understanding of different forms of love. Queer ecologists such as Catriona Sandliands (2010) and Greta Gaard (2011) and queer theorist Jack Halberstam (2020) are asking us to recognize that in different species there are many ways of loving and caring. Sex is not just about the male mating a female to produce offspring. It is about fun, play and care with different partners. Queering sexuality is not only about saying same sex is OK but also about multiple patterns of relations and diverse forms of pleasure in communities. Western science and economics have taught us that heterosexuality is the dominant form, reducing love, desire and forms of care for others into narrow heterosexual norms, fenced in by cultural and legal restrictions. In addition, Indigenous writers, such as Kim TallBear (2021) and Aileen Moreton-Robinson (2021), show how heteronormativity has been imposed through colonisation, suppressing the traditional connections between the productive and reproductive sphere of land use. Indigenous women have mobilised politically to protect and care for the land, resisting colonial logics which disassociate people's connectedness to the earth. There are multiple genders and multiple ways of caring for others in communities which are intergenerational and interspecies.

Queer ecology, then, offers us the possibility to consider different ways of organising and thinking about how to care which go beyond heterosexual relationships and nuclear families to embrace more-than-human relations. However, in terms of the politics of care, challenging restrictive dominant western norms is difficult. Engaging in non-binary ways of thinking means not othering bodies because they don't have a particular physical mobility, or sight, or hearing. Crip and queer studies recognise there are many forms of bodies. They challenge us to consider the norms and ensure that knowledge is produced in ways that recognize bodily diversities.

It is difficult politics even within feminism. There are some feminists who do not recognise trans women as women. So, in terms of care, it is important to build culturally safe, secure spaces, where we can learn and listen and build awareness of what it means to "other". People are going to make lots of mistakes because we're brought up in a certain way of understanding, but we need to hear and learn from diversity. This also requires that we allow space for self-determination and demands for autonomy. It is important to have spaces where people who have been marginalised or othered come in and create trust with each other; spaces where allies are *invited* in—or not.

Marlene: I totally agree, we all need to go deep in this discussion and learn from diversity and the contestation of our privileges trying to relate carefully with *otherwise* meanings (outside dominant racialised Eurocentric norms and cis-hetero, able-bodied norms).

Eoin: I was thinking of how the concept of "otherwise" is one of the central tenets of feminist political ecology, and how, as a network, WEGO has been practising care and *otherwise* knowledges, whether we are thinking *otherwise*, doing *otherwise* and imagining futures *otherwise*. This strikes me as very compatible with the works of theorists such as Sara Ahmed in her thoughts on queer phenomenology. Ahmed (2006) argues that queerness disrupts and reorders social relations by not following accepted paths. This queering or disorientation can put other objects within reach that might not have at first glance seemed possible. I was wondering what strikes you as an exciting challenge for feminist political ecology and practices of care when engaging queer theory, or queer ecology, in your own work?

Wendy: Thank you for bringing in the question of the *otherwise*, which I have learnt from reading and listening to decolonial feminism, as well as in the ideas of transgression or disruption central to queer theory. It is exciting to consider how *otherwise* acknowledges the analytical importance of emotions and love, two concepts which have been traditionally placed outside of academic study. I consider that the emotions of grief and loss are at the centre of responses to climate crises, the COVID pandemic and wars. The person who introduced me to queer ecology was Giovanna Di Chiro (a member of WEGO and long-time friend). Di Chiro's essay on queering environmental discourse on toxic waters and Catriona Sandiland's understanding of death and loss around AIDS were eye opening to me about how to analyse emotions otherwise because their work allows us to look at the validity of feeling grief and fear (Di Chiro, 2010; Sandilands, 2010). The notion of *otherwise* also goes back to the importance of storytelling. The power of poetry, visual arts, theatre and film are all just as valid as academic texts in understanding how we care in the world. Queer ecologies invite us to see the validity of imaginative and speculative forms of knowledge (Haraway, 2016).

COVID has also disrupted our sense of the world, both in terms of its embodied impact and how the pandemic changed our way of relating to each other. The pandemic forced us to recognise how important care for humans and more-than-humans is for our overall well-being, as well as how we have come to exist and relate through a global digital world.

We've learned from that painful time that there is something possible that can emerge in all those online conversations. To me that's also queering, disrupting the normal hierarchy of the academic text over other forms of learning via experience. All of these video calls we've been involved in have become part of the way we relate to the world too. While that definitely needs more unpacking in terms of who profited from our digital connectivity and who was included or excluded, global virtual technology allowed us to connect and relate during Covid. Lastly, through *otherwise* conversations, through feminist and queer intervention, we now bring our passions, loves and fears into the academic world. When I was a student, my sexuality, emotions and fears could not be brought into my work. Now there is space to do so, even if it can be painful, and that to me is an important step towards queering knowledge and creating different forms of ecologies.

Practices of Care in Academe

Eoin: That was a really lovely intervention. I especially like what you said at the end-that for something to be exciting it doesn't have to be this kind of joyful kind of engagement from the very start, it can germinate through painful experiences. I'm wondering what role you think academia can play in opening up space for nurturing practices of care. There seems to be an obvious paradox. On the one hand, the assumption is that the academy is a force for good; on the other hand, it is also part of the problem, given the impact of the neoliberal university, the current mental health crisis among graduate students and the challenges faced by first-generation scholars, who are disproportionately Black students.

Khayaat: I think those are very important considerations, especially if we are to meet the needs of first-generation students and students with different abilities. And I want to pick up on what Wendy was saying about queer ecology. She raises the notion of disruption as a way in which we understand the world, the way in which we engage with the world. I would point to the disruption of hierarchies which is necessary for a care-full university or higher education system. One of the hierarchies is that of age and the assumption that the person who enters the higher education system is a blank slate with no previous knowledge or culture. The kind of university I would like to see would open up and invert hierarchies in order to disrupt the boundaries or the constraints of who belongs in the university and who does not. In the context of the South, universities are for a very select group of people who make it through a not very supportive secondary education system and then find themselves at a university.

We need to bring the university closer to those who cannot access it through traditional meritocracy. Coming back to Wendy's point about stories and about recognition of the valuable caring activities which already occur in the world, we need to learn from such *otherwise* visions and aspirations of communities and social movements. We need to bring those aspirations and stories about care much more into the university. I would like to see the university as much more accessible. In our South African context, just hosting a workshop with a social movement at a university venue is quite significant for those movements. It is a recognition by the tertiary education system that the work of the movement is relevant. But the neoliberal university demands financial gain for every activity, so it can be difficult to do such events, important as they are, because movements cannot afford the venue hire.

At a personal level it is empowering for me to be in a classroom and say, "I don't know everything, but let's come together and let's talk about what we can all contribute to this space", because it is scary to be up there as the all-knowing lecturer. So, for me, those disruptions to hierarchy are exciting, and it is tremendously enriching as a so-called scholar or academic to also be learning all the time. We need to think about our pedagogies, how we value and recognize *otherwise* activities, in order to develop a system of higher education which is more respectful and protective of human dignity, the dignity of its academic members, the dignity of project, administrative and cleaning staff, and students.

If we recognise the cultural differences between first-generation students, the differences of people from different racial backgrounds, we could construct the architecture of our learning spaces as more inclusive. I was shocked to see that people with disabilities cannot access the ISS in the front entrance. There is a system of asking for the front desk to help with a moveable ramp or to go to the back of the building. This is stigmatising differently abled bodies. We also need to think about the people that do the provisioning and cleaning in a university. They are often racialised people. I see that also in my hotel in The Hague. For universities to be caring places we need to reflect where our students are from in the architecture, the food, the accommodation, as well as in our pedagogy and in our learning materials. Similarly, curricula need to reference many different kinds of writing and resources. A question we could ask ourselves is, how much does what we teach link with the experiences of our students? Is there a proper representation of women writers, of Black women writers from different places in the world?

Wendy: To follow on from Khayaat's important observations of how care should be expressed in universities, I would like to refer to my experience of teaching in ISS, a postgraduate school with people from all over the world. Students are mostly people of privilege: even those who access scholarships from rural institutions in the Global South. However, coming to a European institution is unsettling and difficult, especially for students who are newly or differently racialised in the Dutch context. Over the 10 years I have been at ISS, I see how hard it is to be open and create a safe space for people from very different contexts to learn together. There is a major challenge to move beyond hierarchies both by students and colleagues. I was very happy to disrupt hierarchies including my own position as a white female professor.

I would like to mention three strategies which I think make universities more mindful of care. The first is mentoring. I've been mentored by people of all ages when I joined academia as an older person at 52. These are people who have really helped me work through the difficulties of the university. Secondly is safety. My battle as a student was about sexual harassment when the issue of rape was taboo. I continue to be deeply shocked and concerned that universities still overlook instances of sexual harassment and bullying of all sorts with very little systemic response. How can you be learning when you're not safe? Lastly, the role of reframing leadership as really allyship. It is important that older (white) "experts" step aside and become allies. I know that's a different way of looking at leadership, to allow for younger people or for people from the margins to come in and change the system. That is a constructive *otherwise* disruption. Professors should not always be at the front but should step back to give other people space. That sort of allyship means working together collectively. Some of it is happening. I was just listening to a law seminar at the University of Sydney which has changed their curriculum to bring in Indigenous law systems. I want to keep positive about these changes; I don't want to be all gloom. So even if ISS is embedded in the neoliberal structure, it is still a small place which is trying to do things differently. I am glad to be part of that collaboration.

Generational Differences in Spaces for Co-learning

Anna: You both beautifully expressed how you position yourself within academia and how you see universities as spaces for co-learning. I was wondering if you feel a generational difference in how your students care for themselves, with others, with the world, in this context of multiple crises?

Khaayat: I think that there is a generational difference. The younger generation is keen on disruptions. We talk about "doom scrolling" for instance, but among students of a younger generation it is a way of homing in on issues which are discussed through social media and can be very progressive. What I especially enjoy about the younger generation is how they are transforming politics around the body. For example, the discussion about deciding not to procreate, not to have children; that is much more open now. The whole notion of body shaming, or slut shaming, is also recognised. Students and the younger generation in general are much more eloquent and articulate. I feel that I learn from them. I agree with Wendy that we need to focus on what we can build. It is hopeful that younger people are willing to take on boomer-millennial debates, pushing back on what has come before and disentangling it. Why is it that people of a certain age, of 25–30 years old, cannot afford homes in the same way as people could 30 years ago? This issue has implications for care, since giving quality care is made more difficult without appropriate shelter and housing.

I want to comment on allyship. I think it's very important in terms of our political activism and our political existence. We have people with very different experiences and whether we're talking about sexual orientation, physical and mental abilities, race, class and so on, I think we need to take seriously what Wendy has said about allyship. Being an ally sometimes means just listening, and not trying to conceptualise or reframe what the person is saying and to understand that sometimes we have nothing to say, we have nothing to contribute, but we can listen.

Wendy: There are generational differences, and I am always learning from young people. It is not just because you have more years that you are by default wiser. For example, younger people deal with body politics and with leadership differently than I did when I was young. I feel that progressive people in my generation opted out of mainstream politics, and we left the system to the people that have then made it even more of a mess. I regret that choice. I realise now we are the system, and we do need to be part of it and be responsible for it. I am sad personally when students don't challenge me when I give lectures, and I hear later that people were too afraid to ask me because I was the professor. And I always think oh, but how could someone be scared of me? Then I recognize that it is about hierarchies- in universities they are difficult to disrupt and I need to be creative and find other ways to be available.

Returning to the Word "Care"

Anna: Our last question is about the word care itself. Care certainly has become more and more prominent both in academic discussions and activist circles, and more recently so in the context of COVID-19, when people were clapping for caregivers and there was a moment of hope that maybe the world would give much more value to caring practices. However, people—and institutions—care in many different and even conflicting ways. I find the word care quite fascinating because it embraces so many connotations. And it can even have meanings that conflict totally with what we would call a feminist ethics of care, if we consider for example how anti-abortionists care deeply about the foetus or white supremacists care deeply about the white race. We also had very interesting discussions within WEGO with some people in our network feeling much more comfortable with care as a concept than others. Probably there is no one single definition or set of values attached to care, and I think it became very clear during this conversation that there are so many ways to approach it both theoretically and conceptually but also in daily caring practices. Do you sometimes wonder if 'care' is too broad a word that might hold many contentious meanings and runs the risk of being mainstreamed and "softwashed" or co-opted by agendas opposed to what we might consider a feminist ethic and practice of care? Or do you rather say this is precisely why we take a stance to claim a feminist understanding of care—holding its inherent contradictions and difficult questions, such as how our caring might create unequal power relations and exclusions on its own?

Wendy: We're susceptible to social media and what the media says is important to care about, whether it is to consume, or to protest or to opt out and binge watch (I am thinking here how Netflix became such a fixture in my friends and family during COVID confinements—it felt as much a part of self-care as doing indoor exercise). I think there is an evolving understanding of care by feminists particularly around the ethics of care which has expanded from counting care work in the economy to including relations of care with more-than-human others. In her work *Matters of Care* (2017), María Puig de la Bellacasa goes deeply into how care impacts the world on all levels. But there are many and various discussions as we have shown in this conversation. In WEGO, we've managed to collectively consider different meanings of care in a feminist network. We have had (mostly on-line due to COVID) discussions where people had different responses. Some felt awkward about the word care, thinking it is fluffy. But others saw it as a core value to our way of working together in terms of supporting others and doing ethical research otherwise. So, I see feminists as exploring the politics of care in diverse and interesting ways academically and in practice.

Khayaat: There are contradictions in how feminists think about care, and we work with those contradictions in terms of care exactly so that it doesn't get soft-washed just because care is often seen as a very feminised activity. Care can be seen as demeaning by dominant systems because it is mostly done by women who are essentialised as having the "natural" ability to care. It is exciting to work with the contradictions by putting them on the table, rather than trying to get a unified or unifying understanding of what care is. And that's exciting for me, because as we debate the content of care, we can be more inclusive and push theoretical boundaries.

Marlene: Talking about contradictions and going back to the question of age, I agree that care is generational, but it is also intersectional and relational. Care is fluid, it changes all the time, and it is related to the porosity of our bodies that are influenced by imaginaries, experiences and affections we live in our everyday life. Care practices change over time-we might be careless in one period and careful in another. I have seen it with my grandfather and my grandmother who have taught me so many things about caring for the environment, taking care of water, nature and recycling as much as possible. And then I see my mom caring now for the environment more rigorously when before she was a bit careless. So, we care in different ways related to the concerns we live and experience in our everyday life. Thank you for the discussion, what we have shown is how care is a very complex activity embedded in contradictions that we need to continue to question in our research contributions, activism, and everyday lives.

Conclusion

In our intergenerational conversation we discussed the different ways we relate to care for humans and more-than-humans by decentring anthropocentric hierarchies and recovering diverse feelings such as grief and fear from which caring practices emerge. We also highlighted how care becomes a powerful political practice to transform our realities when we ask who can care, with whom or what do we care and how do we care. We explore how the practices of care are intergenerational and political, and how valuing care remains an ongoing feminist struggle that requires continual interventions in public and private institutions. Our focus was on how care is understood in the academe, in teaching as well as theory across the generations in order to underline how care is practised in different ways and at different levels; in ethical and moral systems, social development strategies and public policies. Within these practices, we recovered storytelling and doing things *otherwise* as ways to recognize the untold histories and colonial legacies embedded in our passions, loves, and fears that shape our everyday life.

Funding: This chapter was funded by the Wellbeing Ecology Gender and cOmmunities Innovation Training Network (WEGO-ITN) funded by the European Union's Horizon 2020 research and innovation programme under the Marie Sklodowska-Curie grant agreement No. 764908-WEGO 2018-2021.

References

Ahmed, S. (2006). *Queer phenomenology: Orientations, objects, others.* Duke University Press.

Anderson, B. (2021). Capability, care, and personal assistance: Making connections. *Open Gender Journal, 5.* https://doi.org/10.17169/ogj.2021.109

Bauhardt, C., & Harcourt, W. (Eds.). (2020). *Feminist political ecology and the economics of care: In search of economic alternatives.* Routledge.

Cock, J. (2021). Transformation from below: Questions from the Mabola struggle to achieve a just transition in South Africa. *Transformation: Critical Perspectives on Southern Africa, 107*(1), 1–25.

Di Chiro, G. (2010). Polluted politics? Confronting toxic discourse, sex panic, and eco-normativity. In C. Mortimer-Sandilands & B. Erikson (Eds.), *Queer ecologies: Sex, nature, politics, desire* (pp. 199–230). Indiana University Press.

Escobar, A. (2020). *Pluriversal politics: The real and the possible.* Duke University Press.

Folbre, N. (1994). *Who pays for the kids?* Routledge.
Folbre, N. (1995). "Holding hands at midnight": The paradox of caring labor. *Feminist Economics, 1*(1), 73–92.
Folbre, N. (2014). The care economy in Africa: Subsistence production and unpaid care. *Journal of African Economies, 23*(suppl_1), 128–156.
Gaard, G. (2011). Ecofeminism revisited: Rejecting essentialism and re-placing species in a material feminist environmentalism. *Feminist Formations, 23*(2), 26–53.
Halberstam, J. (2020). *Wild things: The disorder of desire.* Duke University Press.
Haraway, D. (2016). *Staying with the Trouble.* Duke University Press.
Harcourt, W. (2009). *Body politics: Critical debates in gender and development.* Zed Books.
Harcourt, W. (2017). Gender and sustainable livelihoods: Linking gendered experiences of environment, community and self. *Agriculture and Human Values, 34*, 1007–1019.
Heymann, J. (2007). *Forgotten families: Ending the growing crisis confronting children and working parents in the global economy.* Oxford University Press.
Moreton-Robinson, A. (2021). *Talkin' up to the white woman.* University of Minnesota Press.
Puig de la Bellacasa, M. (2017). *Matters of care: Speculative ethics in more than human worlds.* University of Minnesota Press.
Räthzel, N., & Uzzell, D. (2019). Environmental policies and the reproduction of business as usual: How does it work? *Capitalism Nature Socialism, 30*(1), 120–138.
Razavi, S. (2007). The return to social policy and the persistent neglect of unpaid care. *Development and Change, 38*(3), 377–400.
Razavi, S. (2011a). Rethinking care in a development context: An introduction. *Development and Change, 42*(4), 873–903.
Razavi, S. (2011b, November 8). *Transformative social policy for gender equality.* Paper presented at Innovative Approaches to Social Protection: Addressing the Roots of Social Inequality, Poverty and Vulnerability in Developing and Emerging Countries, Friedrich Ebert Stiftung Expert Conference, Berlin.
Sandilands, C. (2010). Melancholy natures, queer ecologies in mortimer-sandilands. In C. B. Erikson (Ed.), *Queer ecologies: Sex, nature, politics, desire* (pp. 331–358). Indiana University Press.
Seager, J. (1993). *Earth Follies: Coming to feminist terms with the global environmental crisis.* Routledge.
TallBear, K. (2021). *Critical indigenous studies handbook.* Routledge.
Tronto, J. C. (1993). *Moral boundaries: A political argument for an ethic of care.* Routledge.

CHAPTER 8

Caring Communities for Radical Change: What Can Feminist Political Ecology Bring to Degrowth?

Stefania Barca, Giovanna Di Chiro, Wendy Harcourt, Ilenia Iengo, Panagiota Kotsila, Seema Kulkarni, Irene Leonardelli, and Chizu Sato

Introduction

In this chapter, we conduct a conversation building from the 8th International Degrowth Conference held in The Hague in August 2021 with

S. Barca · I. Iengo · P. Kotsila
University of Santiago de Compostela, Santiago de Compostela, Spain

G. Di Chiro
Swarthmore College, Swarthmore, PA, USA
e-mail: gdichir2@swarthmore.edu

W. Harcourt (✉)
International Institute of Social Studies, Erasmus University Rotterdam, The Hague, The Netherlands
e-mail: harcourt@iss.nl

W. Harcourt et al. (eds.), *Contours of Feminist Political Ecology*, Gender, Development and Social Change,
https://doi.org/10.1007/978-3-031-20928-4_8

the theme of "caring communities for radical change".[1] In this introduction we start by explaining the synergies between degrowth and FPE. In Section One "Feminist Political Ecology in Conversation: Caring Communities for Radical Change" we summarise the three interventions made by Giovanna Di Chiro, Stefania Barca, and Seema Kulkarni, who presented their approaches to FPE in one of the conferences' plenary sessions. They reflect on their work on environmental and climate justice, gender, and care in order to suggest different ways that communities of care can be fundamental components of radical socio-ecological change towards degrowth. These contributions come from diverse standpoints of theory and different geographies of work and engagement. All point to the contradictions between endless economic growth and ecological balance and social justice. They also emphasise the importance of care and caring communities in resisting, questioning, and counteracting the structural racial, gender, and wider social inequalities that uphold and are perpetuated by growth-dependent economic systems. In Section Two

[1] This chapter is based on a set of conversations around the feminist political ecology plenary held at the 8th International Degrowth Conference in August 2021. The plenary was the culmination of a series of online and in-person conversations between the authors as we collectively thought about how to bring a feminist political ecology perspective to the 8th degrowth conference through theme of "caring communities for radical change". We were interested in dialoguing with the degrowth community about how feminist theories and practices of care can contribute to the strengthening, building and imagining of communities for radical socio-ecological change. The FPE conversation was the WEGO network's contribution to the degrowth conference along with other conversations around decoloniality and arts and culture. The chapter builds on the legacy of earlier International Degrowth Conferences, specifically the 5th and 6th held in Budapest and Malmö, respectively, when the Feminisms and Degrowth Alliance (FaDA) had its first in-person meetings, followed by a number of online exchanges and initiatives, including a collective statement on Covid-19 (FaDA, 2020).

S. Kulkarni
SOPPECOM, Pune, India

I. Leonardelli
IHE Delft Institute for Water Education, Delft, The Netherlands
e-mail: i.leonardelli@un-ihe.org

C. Sato
Wageningen University, Wageningen, The Netherlands
e-mail: chizu.sato@wur.nl

"Towards a Feminist Political Ecology of Degrowth?", we continue these reflections with other scholar-activists who also contributed to the FPE theme at the Conference. The contributions to this section were made individually, and while the chapter has been reviewed by all the authors, it should be noted that each author has contributed their own point of view. Our discussions argue why degrowth needs to take FPE into account when seeking possibilities for just futures. In Section Three "Conclusion", we identify some shared methodological approaches of FPE as we sum up why care, communities of care, and caring practices for radical change are key to degrowth. We conclude with some points for the degrowth movement to take into account from grounded feminist transformative politics.

Synergies Between Degrowth and FPE

Degrowth is both an academic and activist framework, which seeks alternatives to current patterns of economic and socio-ecological destruction and calls for movement towards a fair and liveable future for all. Degrowth proposes a scaling down of certain sectors and aspects of the economy (such as advertising, the military industrial complex, planned obsolescence of products, and the fossil fuel industry) and supports other kinds of economies and policies (such as shorter working weeks, a care income, economies of renewables, and strong public education and healthcare systems). Degrowth transformations are envisioned as taking place in equitable and democratic ways by strategically addressing privileges so as to avoid abrupt changes that will negatively impact the most vulnerable groups and exacerbate injustice (Chertkovskaya et al., 2019; D'Alisa et al., 2014; Liegey et al., 2020; Paulson et al., 2020). Through interdisciplinary analysis, including with and through social movements and grassroots initiatives, degrowth aims to debunk the idea that continuous economic growth is necessary for modern life and civilisation, and that tweaked versions of "green growth" are capable of addressing the interrelated crises of ecological destruction, social injustice, and climate change. As Schneider and Pope (2020) eloquently summarise "degrowth is not a passive critique but an active project of hope".

However, the central issue of unequal power and privilege is often left unquestioned in degrowth propositions, especially in terms of who needs to or can afford to "degrow", whose claims and demands are represented in the degrowth movement, and how subaltern, racialised, and gendered communities at the global and local levels would be impacted by degrowth initiatives, policies, and approaches (Smith et al., 2021). While these tensions and questions are common in many articulations of alternatives (Argüelles et al., 2017), we feel it is important to interrogate power relationships in emerging imaginaries of radical socio-ecological change like degrowth. In this chapter, we outline ways in which feminist political ecology (FPE) perspectives can contribute to degrowth analysis and political strategy. We aim to strengthen the radical potential of degrowth by problematising and pushing forward the questions of who is recognised in, or can be part of, degrowth communities of practice and thought.

An emerging realm of FPE scholarship focuses on power analyses in degrowth (Paulson, 2017), with particular attention to coloniality (Dengler & Seebacher, 2019), race (Abazeri, 2022; Gilmore, 2013), class and labour (Barca, 2019), and gender (FaDA, 2020; Saave-Harnack et al., 2019). Neera Singh (2019), for example, stresses the need for deeper mutual learning between environmental justice movements of the global South and degrowth, in order to discover common ontological grounds for "other ways of being". She argues that an examination of on-the-ground practices and epistemologies of local communities can help "reconceptualize work and care in a post-production, post-growth world" (Singh, 2019, p. 139). In a similar vein, Padini Nirmal and Dianne Rocheleau (2019) propose a feminist and decolonial perspective on degrowth that is "materially and ecologically rooted and culturally expanded" through practices of "re-rooting and re-commoning" (Nirmal & Rocheleau, 2019, p. 470), what they refer to as "regrowth"—practices of regeneration and collective flourishing that counter the destructive legacies and paradigms of colonialist expansion and capitalist forms of growth.

Feminist economists have been among the first to criticise GDP growth from the point of view of class, gender, and colonial inequalities (Gregoratti & Raphael, 2019; Wichterich, 2014). As scholars and activists from the "Feminisms and Degrowth Alliance" (FaDA) have highlighted, rather than understanding degrowth as just about shrinking the economy, degrowth should be focused on transforming core institutions that govern production and reproduction, inspired by and grounded in feminist traditions (Saave-Harnack et al., 2019). Part of this vision and challenge is to re-situate and re-value *care* at the centre of socio-ecological processes and systems: to treat care as a core common for a liveable future. But what is meant by the term and concept of care? And what kind of caring practices and communities can support a radical social transformation?

FPE seeks to answer these questions in diverse ways. FPE is a convening space of research and ideas where scholars theorise different forms of power and access to resources. FPE understands people as embodied and emotional beings with "complex and shifting relationships to the natural world, embedded in place and shaped by interactions of gender, race, class, caste, culture, age (and so on)", moving towards environmental and social justice (Resurrección, 2017, p. 74). In FPE, the concept of care is theorised in various ways. In their edited volume, for example, Wendy Harcourt and Christine Bauhardt (2018), delineate some main threads of thought around care as the gendered work of social reproduction: care as a form of commoning; care as looking after and providing for human and nonhuman others; and care as interspecies reciprocity and more-than-human relations.

FPE thus highlights the central role of socio-ecological production and reproduction and the labour of care as the foundation of planetary well-being by asking more specifically: in what ways can economies be transformed in terms of provisioning and care while degrowing in a socially just manner? What can we learn, in this regard, from communities who are fighting every day for environmental and social justice, or simply for their own well-being and survival on earth? This learning might well go beyond social reproduction and the care of human families and communities to include the work of "earthcare" (Merchant, 1989) and environmental struggles in the defence of and care for more-than-human others (Barca, 2020b; Fragnito & Tola, 2021). What would this expanded understanding of care, beyond but including social reproduction, mean for a new politics of care commoning? As Chizu Sato and Jozelin Soto Alarcón point out, "community is, by definition, constituted through commoning. It is the process and site of being produced through sharing a property, a practice, or a knowledge" (2019, p. 38). Building on these different understandings of care, we ask: what are caring communities in a post-capitalist, post-growth future and what are the main challenges ahead?

In this chapter we contribute to ongoing discussions on care and communities in degrowth by looking at how societies can be reorganised in ways that promote intersectional justice and the sustainability of life. We argue that care is crucial to social and ecological reproduction as we critically reflect on the experiences of paid versus unpaid, collectivised versus feminised care work in order to build just, sustainable, and convivial societies. We propose these ideas as essential contributions to degrowth debates.

Feminist Political Ecology in Conversation: Caring Communities for Radical Change

One of the starting points of the 8th International Degrowth Conference in August 2021 was that "it is not enough to build a movement; we need to build community". In what follows, Giovanna, Stefania and Seema reflect on how this entry point resonates with their work on environmental and climate justice, gender, and care and in what way this contributes to degrowth debates.

Giovanna Di Chiro: Practising Collective Care—Environmental Justice, Kinship, and Interdependence

Many of the panels and workshops at the Degrowth conference were exploring the core question: what does a *caring, strong, and resilient* "community" look like? I'm interested in the questions of *who* we imagine as members of our community and who we see as partners in co-creating a more caring world. I have learned a lot from the members of environmental justice and Indigenous communities with whom I have studied and collaborated over the past four decades.

One of these important mentors for me is the late Grace Lee Boggs, the Detroit-based revolutionary who, in her 100-year life, after having participated in virtually every major social and environmental justice movement of the twentieth and early twenty-first centuries, concluded that "social movements are born of critical connections rather than critical mass" (2012, p. 35). I think that building critical connections and dependable relationships across differences of all kinds is what fuels strong communities and robust socio-ecological movements, especially in times of escalating uncertainty and danger. But, as Potawatomi scholar Kyle Powys Whyte (2020) has argued, trustworthy and durable relationships have been undermined or destroyed by historical and ongoing colonial and capitalist systems. He writes:

> As kinship-based interdependency declines, climatic disruptions can be experienced as abrupt and escalating because responsiveness becomes hard

> to achieve. For whom do we reach out to as trusted partners for coordinated action?...We must take urgent action to establish or repair kinship relationships. Or else we will not have the interdependence required for responsiveness that prevents harm and violence. (pp. 40–42)

For Whyte, and many others, kinship relationships are the foundation of social and ecological *interdependence*, and when these relationships of interdependence are disrupted, then social, ecological, and climate systems are also disrupted. Caring communities are rooted in the recognition of interdependence and being in "right relationship" with all our kin, with all of our human and other-than-human relatives (Tallbear, 2019, p. 31).

The critical scholarship of activists like Whyte, Tallbear, and many others has taught me that, as we strive to build diverse communities working towards a just transition to more equitable and liveable futures, we need to ensure that our calls for degrowth, decolonisation, environmental justice, and caring communities are not just metaphors (Tuck & Yang, 2012). Part of seeing decolonisation (or degrowth) as not just a metaphor, and striving to build diverse communities of care, is about taking seriously the histories of the land that we are living on and learning about the long-standing connections to land and place and the sustainable lifeways practised by the people who have lived there for generations. Many Indigenous feminist scholars have argued that "progressive" green movements such as environmentalism, sustainability, and degrowth rely on colonial abstractions/metaphors of "the commons" as the green spaces, land bases, and natural resources that must be protected and preserved for the "common good" (Arvin et al., 2013; Liboiron, 2021; Whyte, 2020). Embracing the idea of "the commons" (including, for example, the common land bases that are needed to build and operate solar and wind farms, ecological agriculture operations, community land trusts, conservation areas, or recycling centres) without an anti/decolonial lens, sustains colonial land relations and "settler futures" since much of what is considered the *commons* consists of unceded Indigenous lands and unacknowledged historical and ongoing dispossession and devastation of Indigenous lives (Liboiron, 2021, p. 36). For many Indigenous environmental justice scholars, therefore, practising "good land relations" is an essential component for creating truly just and caring communities.

An important piece of my own critical practice has been to engage what Nishnaabeg scholar Leanne Betasamosake Simpson (2017) calls "land as pedagogy" by situating my research and teaching in the places, geographies, and community ecologies in which I live. Potawatomi environmental scientist Robin Wall Kimmerer has argued that part of this practice involves "indigenizing" our relationships with our places. She explains that it is "…not to appropriate the culture of indigenous people", but rather one must "live as if we'll be here for the long haul, to take care of the land as if our lives, both spiritual and material, depended on it" (2013). One example of how I strive to build kin-centric relationships living and working in the Philadelphia metropolitan region is participating in a campus-community collaborative I co-founded in 2012 called Serenity Soular. Serenity Soular's mission is to bring solar technology, sustainable community development, and solar jobs training opportunities to residents in North Philadelphia, a majority-Black section of the city suffering from the harms of environmental racism and long-standing economic and social disinvestment. We spell solar s-o-u-l-ar to emphasise our intention to keeping the soul—or our connection to the people and the commitment to environmental justice—at the forefront of our strategies to build a movement towards a "just transition" for the city and beyond (Di Chiro & Rigell, 2018).

Another example of how I imagine "otherwise worlds" grounded in care comes out of my and my students' collaboration with local activist Zulene Mayfield, long-time leader of Chester Residents Concerned for Quality of Life (CRCQL), a 30-year-old grassroots environmental justice organisation located in nearby Chester, Pennsylvania, an impoverished city whose population is 70% African American and that houses the United States' largest waste incinerator. My students and I have been working with Zulene and CRCQL for several years assisting with campaigns to shut down the incinerator and transition towards a more just, cleaner, zero waste economy in the region. Zulene has always expressed the sentiment that it's important for us to see ourselves as living in kinship relations with residents of nearby Chester, despite the fact that our life conditions may be worlds apart. She has told my students: "Chester does not have a Wakanda Shield over the city keeping the pollutants from travelling all over the region. The pollution doesn't just hover over Chester. The rest of our county is also harmed by the incinerator". This kind of kin-centric environmental justice organising is an active, embodied approach for building critical connections; to use Grace Lee

Boggs' words, it means seeing your own body, your own lifeworlds in the action. Such collaborations challenge the culture of individualism so ingrained in Western thinking. I think these critical connections make tangible what is really meant by interdependence and collective care.

Stefania Barca: Communities of Earthcare

The first story that comes to mind when speaking of "communities of care" is that of Praialta Piranheira, the agroforestry settlement in the Brazilian Amazon which inspired my latest book entitled *Forces of Reproduction* (2020a). Praialta is a particular type of protected area—what the Brazilian law calls "extractive reserve", i.e., public land which is set aside from capitalist exploitation and the GDP growth imperative and managed by communities who provision themselves via sustainable extraction of wild fruits, nuts, seafood, and other non-timber-forest-products. Brazil's extractive reserves originated from grassroots struggles for social justice dating back to the 1980s and to the Alliance of Forest Peoples (Barca & Milanez, 2021). I see these as struggles for the interspecies commons, intended here as a political community made up of forest and people that do not see their humanity as separate from the nonhuman, but rather as co-existence and re-existence with it.

This interspecies community includes Indigenous populations, but also the rubber tappers and other racialised people who call themselves "traditional", and that reproduce themselves with a variety of biomes—not only forest but also riverine ecosystems and mangroves. Their livelihood, food sovereignty, and well-being depend on the wealth of their territories, so they take care of these territories with a spirit of both earth and self-care. Zé Cláudio Ribeiro da Silva and Maria Do Espírito Santo, for example, the *extractivistas* whose story inspired my book, made a living via the extraction of the *castanha do Pará* (Brazil-nut), which they collected from the plot assigned to them within the Praialta settlement. Keeping the *Castanheira* alive and healthy, by defending it from illegal cutting and timber trafficking, was their primary preoccupation—their life project. Their idea of interspecies commoning can be heard from their own voices through the beautiful documentary film *Toxic: Amazonia* (Milanez & Loyola 2011; see also Milanez, 2015).

The communities of earthcare that Brazilians call "extractive reserves" are the product of historical agency, of social conflict and struggle, of resistance to capitalist patriarchy and white supremacy, to GDP growth, and to the financialisation of nature over the past few decades. As the caring logic stands opposite to the logic of extraction (in the capitalist and productivist sense of the term), earthcare communities are constantly threatened and targeted with structural violence—both physical and symbolic. Today, many *extractivistas* are forced to leave the reserve and move to urban areas or to plantations to become proletarians—part of the labour force for capitalist growth. The rationale behind this is to break people's caring relationship with each other and with the land—to turn more-than-human communities into individual proletarians and resources awaiting exploitation. These pressures have been constant throughout the entire history of Brazil, and have escalated during Bolsonaro's government, a fascist mix of heteropatriarchy, white supremacy, and climate denialism (Iamamoto et al., 2021). At the time of writing, the Supreme Federal Court of Brazil is deciding over the government's proposal to drastically reduce the recognition of indigenous territories—a move that spurred the largest Indigenous mobilisation in more than 30 years, with 6000 people camping in the capital Brasilia in the summer of 2021.

Feminist political ecology helps us to make sense of the story of Brazil's earthcare communities by highlighting one key component of their struggle: the relevance and value of care work, extending the focus from the domestic realm to that of the land and nonhuman environment. I borrow the term earthcare from white ecofeminist scholar Carolyn Merchant (1980), who wrote about the historically constructed nexus between women and care in Western culture, to make critical sense of women's agency in a number of environmental mobilisations. However, my understanding of earthcare goes beyond the focus on women, and also beyond environmental mobilisation itself. Inspired by Ariel Salleh's concept of meta-industrial labour (2010), and based on the story of Prailata Piracheira, I see earthcare as the labour of environmental reproduction, i.e., "the work of making nonhuman nature fit for human reproduction while also protecting it from exploitation, and securing the conditions for nature's own regeneration, for the needs of present and future generations" (Barca, 2020a, p. 32). Earth-carers keep the world alive, yet their environmental agency goes largely unrecognised in mainstream narratives of the catastrophic earth-system change epoch that scientists have called the Anthropocene. This invisibilisation of earthcare labour, I argue, has to do with the dominant cultural paradigms of capitalist, industrial modernity, a historical formation which identifies modernity with the "forces of production" and with human geo-supremacy. Undoing the geo-supremacy perspective is thus necessary to see communities of earth-carers as part of a larger historical agency that has been of fundamental relevance to the reproduction of earth systems throughout human history.

Environmental reproduction is a feminist concept, insofar as it is based on a de-naturalisation of reproductive work, as well as making visible its social and ecological relevance beyond the domestic/subsistence sphere. It also aims to call attention to the social processes which tend to appropriate this work and subsume it within capitalist or state-productivist political economy, which prioritise GDP growth over both human and nonhuman life. Like women in social reproduction, so racialised, colonised, and/or low-income people, peasant, Indigenous, and Afrodescendant communities, have historically been assigned the role of reproducers of nonhuman nature—what economists and technocrats now call ecosystem services. Interesting contradictions can be observed in this "hidden abode"—borrowing the term from Nancy Fraser (2014)—of nature conservation. Taking mostly place outside of capitalist wage relations, this work is also non-alienated—i.e., it allows people to engage in a direct relationship with nonhuman nature and to reconnect with their species-being. At the same time, the logic of endless capitalist accumulation pushes towards the disappearance of autonomous and subsistence work by all means, including both symbolic and material violence, by financialising and subsuming all forms of care within capitalist relations. Learning from Maria do Espírito Santo, as she described her involvement in the Praialta project (Milanez, 2015), I see this fundamental contradiction of environmental reproduction as the context from which earthcare struggles emerge, as both organised resistance to value extraction and as the daily micropolitics of *re-existing with* nonhuman nature in relations of care. This is why I have proposed to understand earth-carers as "forces of reproduction" (Barca, 2020a), i.e., historical subjects with a counter-hegemonic potential, which finds expression in organised political struggle, locally, nationally, as well as globally (Goodman & Salleh, 2013).

As the 2021 degrowth conference in The Hague demonstrated, the degrowth movement, which emerged from Western critical consciousness of both planetary and social limits to GDP growth, is evolving towards a fuller realisation of the patriarchal and racial/colonial roots of the growth imperative. Together with decolonial scholarship and activism, FPE makes a fundamental contribution to this effort, pushing a re-definition of degrowth politics based on the assumption that "the foundations of the wealth and well-being of the world rest upon the sphere of social reproduction and the labour of care" (FaDA, 2020). If degrowth politics consists in the search for a radical alternative to the hegemonic paradigm of GDP growth as the foundation of wealth and well-being, then earthcare communities must take central place in it.

Seema Kulkarni: Communities of Care as Gendered Struggles for Agency and Survival

As part of the Mahila Kisan Adhikar Manch (MAKAAM, the Forum for Women Farmers Rights) we engage on various issues faced by disenfranchised and exploited women from diverse socio-economic groups working as agricultural, forest, and livestock (migrant) workers and cultivators across 22 states of India. They face the triple burden of exploitation due to caste, gender, and class, yet they exhibit agency in multiple ways—much of it being necessary for survival. MAKAAM voices the concerns raised by its diverse members among whom are various grassroots organisations, women farmers, and labourers.

Lata is a migrant sugarcane worker, who belongs to a disadvantaged caste and class, working in the western Indian State of Maharashtra. Today, sugarcane is used not only for producing sugar. The sugar lobby is increasingly diversifying into multiple trades like ethanol and alcohol production. The sugar factory extracts more and more profit from the cheap labour of the migrant workers, especially women like Lata. Typically, contracts are made with the man of the household, leaving Lata without direct access to her wage. She and her husband work as a unit, referred to as *Koyta* (sickle), and perform arduous tasks, such as harvesting cane, tying cane bundles, and loading, unloading, and transporting them to the factory. The working day is usually 12–13 hours long as the contractor and middleman insist each *Koyta* to meet the daily target of harvesting two tonnes of cane for the factory owner, who aims to crush the cane to the maximum capacity of the factory to maximise profits. Lata and other women workers wake up at 3 am to load the trucks, often at the cost of their health. None can afford illness since missing work incurs a huge fine which is usually twice that of the wages earned. Menstruating and pregnant women thus continue to work despite the discomfort. Some women work till the last hour of their pregnancy and deliver at the worksite itself. Single women workers often suffer sexual harassment at the workplace and have to carry their young children around during work (Shukla & Kulkarni, 2019).

When Lata is not migrating for her survival, she leases land from an upper caste landlord and engages in subsistence farming to support herself and her family. Being a poor woman from a "lower" caste has meant that she rarely receives timely support to carry out her agricultural work, and even the delayed support from the upper caste communities comes only if she can return the favours, for example, sexual favours or exchange of free labour. As a result, Lata not only loses out on getting good yields, but her mental peace is also compromised. Lata loses out because of her caste, class, and gender disadvantage. In addition to the long working day as both a sugarcane migrant worker and a subsistence farmer at home, Lata devotes additional four to five hours for unpaid household work: childcare, cooking, cleaning, and fetching fuel and water. MAKAAM supported individual women sugarcane workers to form the Women Sugarcane Harvesters Organisation, so that Lata and women farmers like her became associated with MAKAAM through their grassroots organisation.

There are many contradictions when it comes to conceptualising or reflecting on notions of care without paying attention to the gendered division of labour. The notion of care depoliticises the question of unpaid work done by women, if it is not grounded in a Marxist feminist understanding that questions capitalist accumulation based on the free labour of women, like Lata, who belong to the disadvantaged sections of society. Women's non-wage work, variously called care work or reproductive, emotional, or affective labour, is necessary for the existence of wage work and for the accumulation of capital. Broadly speaking, mainstream thinking and that from some quarters of environmental groups focus on the economic and ecological crisis, and the solutions (e.g., ecological farming and climate smart agriculture), often gloss over the unpaid or non-wage care work of women. Ecological health and the contributions of women's unpaid care work towards it will thus have to be framed in a manner that addresses upfront the question of women's ownership and control over resources. For example, ecological agriculture calls for caring for soils, selecting and conserving local varieties of seed, bringing a diversity of crops to the farm, or managing an integrated farm with backyard poultry and other animals. It is assumed that this work would be done by the women of the household, while men continue with the business-as-usual commercialised farming that includes the cultivation of sugarcane, cotton, etc. Such harm will have to be addressed by centring the issue of care in development thinking. When considering social and political organising, we need to consider the struggle for entitlements, rights over resources, and women's unpaid work alongside the call for ecological health. By working with MAKAAM I have learned that both are needed and must go hand in hand. Ecological health cannot be framed within an extractive capitalist paradigm which free rides on the non-wage work of women and the poor and the environment.

We demand the state give us as a matter of right not only land and welfare, but also support to rebuild our soils and our lives through agroecological farming. Our two-fold strategy is thus one of reimagining our world that was in harmony with nature while addressing discrimination based on class, caste, and gender. Our struggles, mobilisation, and demands articulate both of these positions. These combined demands call for a just society that cares and values the knowledge and work of the women and men of disadvantaged communities. It values and cares for nature—land, water, and forests.

Drawing on the work of Maria Mies (1998) and others we believe that the solutions lie in reconceptualising the concepts of economy and labour. The question of women's work and rights, and the question of ecological, social, and economic sustainability have to be placed at the centre of our analysis and politics. But this requires a different view of the economy and of society, which requires that we start paying attention to the work of nature and her regenerative cycles, and valuing women's unpaid care work in the household as well as all other non-wage work for subsistence. We also have to keep in mind that communities are not homogenous and cohesive but have diverse groups, often in conflict with each other. The women we are working with have long been exploited due to their class and caste positions in this capitalist world. Their struggles have thus been to fight the caste-based capitalist patriarchy. In this new framing we have to be conscious that the burden of care does not rest with women and the disadvantaged social groups who have thus far carried this burden on their shoulders. Care and social justice will have to go hand in hand if "otherwise desirable worlds" have to become a reality.

Towards a Feminist Political Ecology of Degrowth?

Following these insights on environmental and climate justice, gender, care, and degrowth by Giovanna, Stefania and Seema, we now discuss two challenges for a FPE degrowth, one concerning how to create communities of care as we reckon with our troubling past, and the other problematising normative assumptions about health and well-being from an embodied socio-environmental justice perspective.

Linking Communities of Care and Reckoning with the Past

Panagiota: Stefania spoke of environmental defenders, and of the violence suffered as part of a struggle to care *with* nature—which she expressed through this concept of *florestania*—a violence brought about by political forces who want to break these relationships of care developed between people and more-than-human, these kinships formed in the forests in Brazil. At the same time, most recent work and activism around the commoning of social reproduction (Federici, 2019) and introducing a care income—the idea of recognising and remunerating care work—highlights perhaps more gendered aspects of care as socio-ecological reproduction and brings forward the demand of reclaiming care as commons, in a way of taking common responsibility for it in society (Barca, 2020a, p. 7). What can be potential common points between environmental defenders as caring communities, and the communities and economies that could be formed around this idea of care as commons?

Care can be understood as a "glue" concept that links demands on visibilising and valuing socio-ecological reproductive labour (the work of care and earthcare) and aligns with ecofeminist thought on how the domination and subordination of female—and other racialised, LGBTQI+, Indigenous, lower-caste and ethnic minority or otherwise marginalised—people, parallels the abusive extractivist activities that change environments and threaten life itself. Thus, care can signal the need to put life and the everyday activities that ensure the physical and emotional well-being of people along with ecological well-being, at the centre of politics and the economy (Pérez Orozco, 2014, p. 93).

In this sense, there are communities who engage with different facets of caring interdependencies: those who organise to protect forests, those who struggle for creating agricultural or solar commons, and others who embody and fight against climate and environmental injustice in risky landscapes. A key question which we need to continue to ask is, then, how these different communities and struggles can come together, inform and complement each other.

How can we bring into conversation the everyday realities of care work and the work of doing environmental justice, in a translocal manner, horizontally, and with a common goal to change the paradigm, the narrative, and eventually, the system, towards a more complete horizon of degrowth attuned to socio-environmental justice? How important is it especially now, to speak about histories of violence and oppression as well as of stories and events that are able to inspire hope—at a time when calls for urgent action (e.g., related to climate change and often directed to those in power) in some ways risk obscuring the voices of activists and affected groups that have been *already* acting, struggling, dealing with climate change impacts and devising alternatives?

Seema: Engagement with the past is critical for reimagining the future. Social discrimination and exploitation as a result of caste, class, and capitalist patriarchy all need to be understood in a historical context. Challenging structural inequalities and unequal power relations becomes possible only in the full awareness of the histories of this exploitation. The agenda of justice-focused movements must be to move away from this past and into reimagining a future which is based on equality and equal opportunities. Can degrowth include these unsettling historical dynamics when it engages with and mobilises elements of the past that cared for nature and lived in harmony with it for a mutual co-existence, in its questioning of the growth trajectory and its limits in relation to planetary ecological balance and social justice?

FPE does not lend itself to easy generalisations. It carries within it some of the contradictions produced by patriarchal societies by identifying women with nature, as highlighted by early ecofeminist scholars, such as Mary Mellor (1994) or Carolyn Merchant (1989). Feminists working in the South were concerned with the material survival of women in poor communities who depended on natural resources (Agarwal, 1992; Mies & Shiva, 1993; Shiva, 1989). Diverse positions articulated by feminists in conversation with the degrowth movements need to be acknowledged and contextualised as we reimagine our futures. Women, especially from disadvantaged communities, are likely to be excluded if present material inequalities go unchallenged. In the Indian context for example, the continuation of the caste system is beneficial for the powerful upper caste communities as they can retain control over means of production, division of labour, and knowledge.

Irene: Building on from what Seema states, patriarchal social norms keep marginalising women across caste and class. This is evident in rural India, where the conceptualisation of "farmers" most often reproduces the imaginary of farmers as male landowners (Agarwal, 2003; Padhi, 2012), although women perform most of the everyday work in the farm. For instance, in rural Maharashtra, where I conducted fieldwork in collaboration with Seema and other colleagues working at Soppecom, it is most often women farmers who take care of sowing, weeding, harvesting, and milking cows. Increasingly, they also engage in irrigation, a task traditionally seen as part of the male domain. From morning to evening they have their hands in the soil: their lives, and their bodies are profoundly woven together, or co-become, with the soil and the water that allow them to farm. Despite being marginalised institutionally and socially, I have often appreciated how women farmers have their ways to influence decision-making processes within the household. They play a role in deciding what to cultivate and how to organise the space of the farm; they often organise work in collective ways and collaborate, increasingly across caste lines. Counteracting patriarchal and casteist dynamics is often a subtle process. Yet simultaneously, water is increasingly scarce and of poor quality; climate change makes weather conditions increasingly unpredictable; agricultural costs keep increasing and, with it, indebtedness; the market is very volatile. There is pride and joy, and hope for the future, as well as pain and fear in women farmers' narratives.

Building caring communities, as building kin-centric communities, is a process that must go hand in hand with caring for forests, waters, and farms. It entails thinking about more-than-human ethics as a process of co-becoming across scales as Puig de la Bellacasa (2017) has suggested. To engage in such a process, it is important to learn from communities and organised movements that are *already* striving for radical change, refusing to align themselves to the violent heteropatriarchal-productivist logics of capitalism. Yet along with organised movements and struggles, a way to create transnational feminist caring communities is also that of learning from, giving space, and creating alliances with those who are fighting for more equitable and ethical presents and futures in their everyday life, often in subtle, yet transformative ways. Moving from my position of privilege and recognising the uneven and non-innocent threads through which more-than-humans co-become on earth, I orient my engagement with women farmers in Maharashtra in this direction—inspired by the work of Seema and other members of the MAKAAM network.

Chizu: In this regard, lessons can be learned from FPE, in particular the combination of anti-essentialist Marxism and community economies, which makes it possible to see beyond the capitalist economy in ways that strengthen the project of degrowth. In Marxist thought, exploitation occurs when surplus—labour above what is necessary to reproduce the labourer—is appropriated by non-direct labourers (Gibson-Graham et al., 2000). Exploitation thus occurs not only in capitalist but also in feudal, and slave class processes, when non-direct labourers appropriate direct labourers' surplus. While the existence of capitalist class processes in the forms of capitalist extractivism in the Brazilian Amazon and capital accumulation by sugarcane factories in India is hard to miss in the reflections, there is at least one non-capitalist exploitative class in the reflections: a feudal class process within a household where a husband appropriates surplus from his wife's non-waged care work.

When building communities of care using an anti-essentialist Marxian ethic, it is crucial to recognise that exploitation deprives labourers of opportunities to invest the surplus they produced in building a community in which members take care to meet each other's needs (Community Economies Collective, 2001). Furthermore, more-than-human community economies scholars (e.g., Gibson-Graham & Miller, 2015; Roelvink & Gibson-Graham, 2009; Sato & Soto Alarcón, 2019), illuminate diverse economies performed not only by humans but also by earthothers. These writers point to the interdependence of different species economies. In neoliberal capitalist relations, labour that is productive is the labour that produces more surplus, and more-than-human earthothers are objects to be consumed in production. These three reflections make visible how these relations and practices disrupt kin-centric relationships between humans and earthothers. The combination of anti-essentialist Marxism, community economies, and FPE perspectives offered by Giovanna, Stefania, Seema, and others, enable us to see often unpaid and invisible care work performed by people from marginalised communities and more-than-human earthothers. Once seen, this care work can now be recognised as productive, not because it produces more surplus but because surplus is used to build more convivial communities for all its members of the community, while discouraging socio-ecological relations recognised as exploitative and unjust through a decolonial, intersectional lens.

Looking at degrowth (and degrowing economies) from this perspective forces us to recognise that there is no single economy. Any economy is constituted by a constellation of multiple economies and human economies are nested in human and more-than-human economies. This perspective, articulated in the reflections, redefines exploitation based on an FPE ethic of care, insofar as exploitation deprives both people and earthothers of opportunities to build a more convivial, socially, and environmentally just, kin-centric multispecies community. This understanding offers a framework that is useful for discovering how to work towards degrowth together, while remaining sensitive to historically developed, hierarchical, or unequal socio-ecological relationships that obstruct building a community of degrowth.

Embodied Social and Ecological Health

Ilenia: As Giovanna, Seema and Stefania urge us to reckon with oppressions which have been historically and materially determined by focusing on movements and experiences who strive to dismantle institutionalised violence, our conversation brings into question the issue of futurity.

Degrowth at its core engages with desirable notions of the future. FPE can amplify the standpoint, experiences, and propositions of those who have been absent or even excluded from radical political imaginaries. In the brilliant book entitled *Care Work: Dreaming Disability Justice* Leah Lakshmi Piepzna-Samarasinha (2018) reflects on Qwo-Li Driskill's words when they say, "one way ableism works is that disabled people are not even present within the imaginations of a supposedly radical future" (p. 215). It is time we take up their invitation to look at the absences that exist within degrowth imaginaries of the future. For a very long time, we have missed feminist interventions on the embodied and multifaceted understandings of social and ecological health that complicate the binary thinking which posits and understands human and more-than-human health as the opposite of sickness, disability, unproductiveness, and disposability; in other words, what Giovanna has called in some of her previous work, "ecoheteronormativity" (2010). It is important to produce anti-ableist understandings of socio-ecological health, reflecting on how it changes over time and place, how it is politicised from different positionalities and contexts and put at the centre of multiple tactics of anti-capitalist, anti-racist, anti-heteropatriarchal, and decolonial coalition-building work. By way of shaking up some of the underlying assumptions about ability in degrowth, we bring ageing and disabled bodies and communities into current imaginaries of degrowth futures and provide room for representation of crip and subaltern kinship without exploiting or instrumentalising them as a resource or inspiration. Building on our previous interventions, I would like to bring in Crip theory and disability justice (Clare, 2017; Kafer, 2013; Piepzna-Samarasinha, 2018, see also Chapter 4 in this volume) in order to point out how it is important to see disability through an intersectional lens. Listening to those who have been excluded or oppressed due to disability can help to open the conversation about the assumptions concerning the natural bodies. Unquestioned assumptions around bodies and sexualities need to be questioned in degrowth discourses and practice to give space to dissident/othered embodiments and trouble and transform the futures we envision towards wider and more inclusive justice. We bring to degrowth the insights of feminist science and technology studies in order to challenge normative values that exclude and create barriers around disability.

Given Giovanna, Seema, and Stefania's experience and knowledge in social and environmental justice organising, we move to focus on who is included or excluded in our radical political imaginaries, such as degrowth. What kinds of knowledge, bodies, and territories, are at the margins of inclusion/exclusion not only of our current unjust societies but also of our political/radical organisations and communities?

Stefania: One example of such inclusion/exclusion dynamics relates to occupational health struggles in working-class communities. Occupational health has been long associated with public and with environmental health, but both scientific practices and management and regulation have tended to hide or underplay those connections to separate the three spheres of health (public, occupational, environmental). Moreover, "jobs blackmail" and other ideological constructs have divided communities and obstructed their search for justice and economic alternatives. Here is where the inclusion/exclusion dialectic becomes essential. Taking the male industrial worker as a reference point for healthiness has long misguided the understanding of the effects of industrial hazards on larger communities—including people with different bodies.

The male breadwinner's sacrifice for family, community, and the nation's GDP implies the silent and misrecognised sacrifice of others around them whose bodies were left out of the account. There is a whole history to be written about the sacrificing of women's health to industrial development and their agency in reshaping dominant conceptions of health by enlarging the boundaries of whose health is accounted for, and what health means in the first place. Children's health, or that of nonhuman animals and life-support systems to which working-class communities are linked through relations of care and interdependency, are similarly excluded.

In some cases, however, the affected people themselves, and the social movements they have built, have struggled for recognition of the non-separation between different kinds of bodies and their environment. Mobilisations against the Ilva steel plant in Taranto, Italy (Barca & Leonardi, 2018) testify to a specific kind of environmental justice, what I have called working-class environmentalism: struggles for reproduction led by working-class communities in recognition of the fact that industrial growth was built upon the *sacrifice*—the supposed disposability—of their bodies and of nonhuman life in their territories.

Giovanna: A core body of critical ecofeminist literature and activism joins together disability theory, queer theory, and environmental justice praxis articulating the values of collective care and kinship relations. The work of scholar-activists such as Patty Berne and Vanessa Raditz (2019), Eli Clare (2009), Shayda Kafai (2021), Alison Kafer (2013), Mia Mingus (2022), and Leah Lakshmi Piepzna-Samarasinha (2021) highlights how people living with disabilities and queer people know how to build caring communities for mutual thriving in the face of multiple crises because they have always had to "crip the apocalypse" in order to survive in a world that wants them disappeared or dead. Creativity and collective care are the cornerstones for "dreaming wild disability justice futures", argues Piepzna-Samarasinha (2021, p. 250). Such futures are made possible through the creation of "vibrant, innovative, crip-made forms of organising" such as new collectivities, alternative care-based institutions, and accessibility hubs and "homespaces" that are "continuing to save everyone's asses" providing sustenance, shelter, personal protective equipment (masks), legal assistance, and mutual aid in response to the violence of escalating climate change, global pandemics, and social precarity (ibid., p. 254).

Refusing pain—or "damage-centered" theorising and organising (Tuck, 2009), these scholar-activists are foregrounding how marginalised communities are challenging capitalism's fantasy of the "self-made man", a fantasy based on the cultural illusions of individualism and self-sufficiency that can only be sustained by devaluing and making invisible the essential social reproductive labour performed by women, working-class people, and racialised communities (see Barca, 2020b; Sato & Soto Alarcón, 2019). Instead, these writers document how queer, crip, BIPOC, and low-income communities are reimagining and prefiguring climate resilience and flourishing lifeworlds by building anti-colonial, post-capitalist economies of repair and care grounded in the principles of "interdependence and collectivity" (Clare, 2009, p. 106). Echoing this point, long-time feminist, disability, and queer activists Patty Berne and Vanessa Raditz (2019) write that we must "see parallels in the havoc that capitalism and the drive to hoard wealth has wreaked on our bodies as queer people, gender nonconforming folks, and people from colonized lands, and how capitalism has abused and exploits the land". Arguing that our collective futures depend on our recognition of the importance of caring for and sustaining both human diversity and biodiversity, they continue:

> The forces of capitalism, racism, ableism, transphobia, and homophobia may have cornered us into a vulnerable position in this unprecedented moment in our planet's history, but the wisdom we've gained along the way could allow us all to survive in the face of climate chaos. The history of disabled queer and trans people has continually been one of creative problem-solving within a society that refuses to center our needs. If we can build an intersectional climate justice movement—one that incorporates disability justice, that centers disabled people of color and queer and gender nonconforming folks with disabilities—our species might have a chance to survive.

Wendy: Reflecting on bodies and health in this discussion, I would like to add my reflections here about embodying degrowth. The idea of growth reproduces its hegemony through everyday practices and performances. In unsettling notions of gender, race, heteronormativity, and able-bodiedness we need to delink from the everyday invisibilising of difference. In the invitation to notice the everyday and embody degrowth, FPE takes up feminist decolonial meanings of seeing the body as a place from where we can start telling stories and find inner strength. As Ilenia indicates, FPE invites cultural and political resistance to the dominant patriarchal (medicalised and racialised) understanding of the "normal" body as white, male, Western, and heterosexual from which all "other" forms of bodies differ. This form of body politics opens a space for transformative collective action which connects the body to radical alternatives offered by degrowth, asking that radical change be founded on diversity and the need for care of our own and other bodies, paying attention to the silencing of difference.

As Stefania and Giovanna have underlined, FPE further invites degrowth to be open to the possibility of talking about *other worlds* and about knowledge *otherwise* in order to unsettle dominant views of what it is to be human from world views outside the colonial frame. We need to understand how to work both across and outside a colonial frame; what are the possibilities of undoing and unsettling—*not replacing or occupying*—Euro-American conceptions of what it means to be human. What would it mean for degrowth to take on black feminist and science studies concepts of unsettling in order to constructively shift Eurocentric positions of degrowth?

In these invitations, FPE is looking at ways of relating, undoing the imaginary of growth through everyday practices. Learning from decolonial and Indigenous feminisms, ideas of relationality, responsibility, and conviviality, and walking with others in allyship, FPE's invitation is to move towards a resurgence of other ways of doing and thinking. FPE invites an openness to a plurality of perspectives and the resurgence of resistances through degrowth. Together, FPE and degrowth can build a shared, pluriversal project, capable of being home to diverse knowledge, languages, memories, and perspectives.

Conclusion

Using the 8th International Degrowth Conference as a springboard, in this chapter we explored the themes of care and caring communities and radical change from FPE perspectives and how they contribute to degrowth debates. As our chapter indicates, FPE perspectives are informed by diverse and, at times, conflicting theoretical approaches. What unites these diverse perspectives are methodological choices such as learning with marginalised communities, valuing their struggles for collective well-being, recognising more-than-human earthothers as kin in collective survival, and looking at context-specific stories as analytical starting points. We share a relational ontology and belief in the importance of intersectionality in shaping environmentally just futures. In conclusion we look below at how our considerations for care and caring communities can strengthen degrowth's transformative potential.

Our focus in this chapter has been on FPE understandings on care rooted in our respective experiences with communities of place-based struggles for social and environmental justice. Care is critically recognised as central to both social and ecological reproduction of humans, economies, and lifeworlds. What we have pointed to is the radical potential of care collectively performed by humans with more-than-human earthothers, in work that is non-hierarchically organised among species and experienced democratically among communities. This care is a kind of glue, binding species across differences of all kinds constituted by kin-centric relationships in place.

Our chapter pointed to examples of such care—Serenity Soular and CRCQL in the US, earth-carers in the Brazilian Amazon and the MAKAAM in India. These stories illustrate what caring communities for a post-capitalist, post-growth present and future look like. Caring communities challenge the culture of individualism, undoing the geo-supremacy perspective, and while still working within states. We have pointed to how caring practices support radical social transformation including degrowth by forming strong kin-centric relationships. They also develop an ability to value "forces of reproduction", which are unseen and devalued in the capitalist, productivist paradigm (such as unpaid care work performed by women and others in the marginalised communities). They also address the discriminations experienced by marginalised communities and the harm experienced by the environment. Caring practices demand people and their more-than-human kin entitlements and rights to resources and ecological health and well-being. We have argued that taking note of such caring practices is not about romanticising "a community" and "the commons", but taking seriously the histories of land and place and learning from the peoples and earthothers lifeworlds that nourish and sustain each other over generations.

Our chapter points to the need for degrowth to take into account caring practices aware of who and what is excluded from transformative politics. We have pointed out the importance of the embodied experiences of marginalised groups: people with disabilities, queer and gender nonconforming people, people of colour, women performing industrial wage labour in working-class communities, their children and nonhuman animals and life-support systems. We ask that these lives are not marginal but central to transformative politics. Degrowth needs to take into account decolonial feminisms, crip theory and anti-essentialist Marxism as a necessary condition for democratic politics. Radical concepts such as "commoning", "community" need to be constantly elaborated to ensure decolonial, non-capitalocentric perspectives are at the heart of degrowth critiques and transformative practices. Finally, we must pay more attention to our own bodies and our own lifeworlds and their interactions. By performing care practices together, in place and time, and making critical connections with earthothers we can produce stronger communities and movements for radical change.

Funding: This chapter was funded by the Wellbeing Ecology Gender and cOmmunities Innovation Training Network (WEGO-ITN) funded by the European Union's Horizon 2020 research and innovation programme under the Marie Sklodowska-Curie grant agreement No. 764908-WEGO 2018-2021.

References

Abazeri, M. (2022). Decolonial feminisms and degrowth. *Futures*, *136*(1) (February), 102902. https://doi.org/10.1016/j.futures.2022.102902

Agarwal, B. (1992). The gender and environment debate: Lessons from India. *Feminist Studies, 18*(1), 119–158. https://doi.org/10.2307/3178217

Agarwal, B. (2003). Gender and land rights revisited: Exploring new prospects via the state, family and market. *Journal of Agrarian Change, 3*(1–2), 184–224. https://doi.org/10.1111/1471-0366.00054

Argüelles, L., Anguelovski, I., & Dinnie, E. (2017). Power and privilege in alternative civic practices: Examining imaginaries of change and embedded rationalities in community economies. *Geoforum, 86*(November), 30–41. https://doi.org/10.1016/j.geoforum.2017.08.013

Arvin, M., Tuck, E., & Morrill, A. (2013). Decolonizing feminism: Challenging connections between settler colonialism and heteropatriarchy. *Feminist Formations, 25*(1), 8–34.

Barca, S. (2019). 'An alternative worth fighting for': Degrowth and liberation of work. In E. Chertkovskaya, A. Paulsson, & S. Barca (Eds.), *Towards a political economy of degrowth* (pp. 175–192). Rowman & Littlefield.

Barca, S. (2020a). *Forces of reproduction: Notes for a counter-hegemonic Anthropocene.* Cambridge University Press.

Barca, S. (2020b, April 7). Within and beyond the pandemic: Demanding a care income and a feminist green new deal for Europe. *Undisciplined*

Environments. https://undisciplinedenvironments.org/2020/04/07/within-and-beyond-the-pandemic-demanding-a-care-income-and-a-feminist-green-new-deal-for-europe/

Barca, S., & Leonardi, E. (2018). Working-class ecology and union politics: A conceptual topology. *Globalizations, 15*(4), 487–503. https://doi.org/10.1080/14747731.2018.1454672

Barca, S., & Milanez, F. (2021). Labouring the commons. Amazonia's extractive reserves and the legacy of Chico Mendes. In N. Räthzel, D. Stevis, & D. Uzzell (Eds.), *Handbook of environmental labour studies* (pp. 319–338). Palgrave.

Berne, P., & Raditz, V. (2019, July 31). To survive climate catastrophe, look to queer and disabled folks. *Yes! Magazine*. https://www.yesmagazine.org/opinion/2019/07/31/climate-change-queer-disabled-organizers

Boggs, G. L. (2012). *The next American revolution: Sustainable activism for the twenty-first century*. University of California Press.

Chertkovskaya, E., Paulsson, A., & Barca, S. (Eds.). (2019). *Towards a political economy of degrowth*. Rowman & Littlefield.

Clare, E. (2009). *Exile and pride: Disability, queerness, and liberation*. South End Press.

Clare, E. (2017). *Brilliant imperfection: Grappling with cure*. Duke University Press.

Community Economies Collective. (2001). Imagining and enacting noncapitalist futures. *Socialist Review, 28*(3/4), 93–135.

D'Alisa, G., Demaria, F., & Kallis, G. (2014). *Degrowth: A vocabulary for a new era*. Routledge.

Dengler, C., & Seebacher, L. M. (2019). What about the global South? Towards a feminist decolonial degrowth approach. *Ecological Economics, 157*(March), 246–252. https://doi.org/10.1016/j.ecolecon.2018.11.019

Di Chiro, G. (2010). Polluted politics? Confronting toxic discourse, sex panic, and eco normativity. In C. Mortimer-Sandilands & B. Erickson (Eds.), *Queer ecologies: Sex, nature, politics, desire* (pp. 199–230). Indiana University Press.

Di Chiro, G., & Rigell, L. (2018). Situating sustainability against displacement: Building campus-community collaboratives for environmental justice from the ground up. In J. Sze (Ed.), *Sustainability: Approaches to environmental justice and social power* (pp. 76–101). NYU Press.

FaDA. (2020). *Feminist degrowth reflections on Covid-19 and the politics of social peproduction*. Degrowth.info. Retrieved March 29, 2022 from https://www.degrowth.info/blog/feminist-degrowth-collaborative-fada-reflections-on-the-covid-19-pandemic-and-the-politics-of-social-reproduction

Federici, S. (2019). *Re-enchanting the world: Feminism and the politics of the commons*. Pm Press.

Fragnito, M., & Tola, M. (2021). *Ecologie della cura: Prospettive transfemministe*. Orthotes Editrice.

Fraser, N. (2014). Behind Marx's hidden abode. *New Left Review, 86* (March/April), 55–72.

Gibson-Graham, J. K., & Miller, E. (2015). Economy as ecological livelihood. In K. Gibson, D. B. Rose, & R. Fincher (Eds.), *Manifesto for living in the anthropocene* (pp. 7–16). Punctum books.

Gibson-Graham, J. K., Resnick, S. A., & Wolff, R. D. (2000). Class in a post-structural frame. In J. K. Gibson-Graham, S. A. Resnick, & R. D. Wolff (Eds.), *Class and its others* (pp. 1–22). University of Minnesota Press.

Gilmore, B. (2013). The world is yours: 'Degrowth', racial inequality and sustainability. *Sustainability, 5*(3), 1282–1303. https://doi.org/10.3390/su5031282

Goodman, J., & Salleh, A. (2013). The 'green economy': Class hegemony and counter-hegemony. *Globalizations, 10*(3), 411–424.

Gregoratti, C., & Raphael, R. (2019). Maria Mies's and Marilyn Waring's critiques of growth. In E. Chertkovskaya, A. Paulsson, & B. Stefania (Eds.), *Towards a political economy of degrowth* (pp. 83–98). Rowman & Littlefield.

Harcourt, W., & Bauhardt, C. (2018). Introduction: Conversations on care in feminist political economy and ecology. In W. Harcourt & C. Bauhardt (Eds.), *Feminist political ecology and the economics of care* (pp. 1–15). Routledge.

Iamamoto, S. A. S., Mano, M. K., & Summa, R. (2021). Brazilian far-right neoliberal nationalism: Family, anti-communism and the myth of racial democracy. *Globalizations*, 1–17. https://doi.org/10.1080/14747731.2021.1991745

Kafai, S. (2021). *Crip kinship: The disability justice & art activism of sins invalid*. Arsenal Pulp Press.

Kafer, A. (2013). *Feminist, queer, crip*. Indiana University Press.

Kimmerer, R. W. (2013). The rights of the land. *Orion Magazine*. https://orionmagazine.org/article/the-rights-of-the-land/

Liboiron, M. (2021). *Pollution is colonialism*. Duke University Press. https://doi.org/10.1515/9781478021445

Liegey, V., Nelson, A., & Hickel, J. (2020). *Exploring degrowth: A critical guide*. Pluto Press.

Loyola, B., & Milanez, F. (Directors). (2011). *Toxic: Amazon* [Film]. VICE Media Inc.

Mellor, M. (1994). Varieties of ecofeminism. *Capitalism Nature Socialism, 5*(4), 117–125. https://doi.org/10.1080/10455759409358615

Merchant, C. (1980). *The death of nature: Women, ecology, and the scientific revolution*. Harper & Row.

Merchant, C. (1989). *Eaerthcare: Women and the environment*. Routledge.

Mies, M. (1998). Globalization of the economy and women's work in a sustainable society. *Gender, Technology and Development, 2*(1), 3–37. https://doi.org/10.1177/097185249800200101

Mies, M., & Shiva, V. (1993). *Ecofeminism.* Zed books.

Milanez, F. (2015). *'A ousadia de conviver com a floresta': Uma ecologia política do extrativismona Amazônia* (Doctoral dissertation, University of Coimbra). https://estudogeral.sib.uc.pt/handle/10316/29762

Mingus, M. (2022, January 16). *You are not entitled to our deaths: COVID, abled supremacy & interdependence.* Leaving Evidence. https://leavingevidence.wordpress.com/2022/01/16/you-are-not-entitled-to-our-deaths-covid-abled-supremacy-interdependence/

Nirmal, P., & Rocheleau, D. (2019). Decolonizing degrowth in the post-development convergence: Questions, experiences, and proposals from two Indigenous territories. *Environment and Planning E: Nature and Space, 2*(3), 465–492. https://doi.org/10.1177/2514848618819478

Padhi, R. (2012). *Those who did not die: Impact of the agrarian crisis on women in Punjab.* Sage.

Paulson, S. (2017). Degrowth: Culture, power and change. *Journal of Political Ecology, 24*(1), 425–448.

Paulson, S., D'Alisa, G., Demaria, F., & Kallis, G. (2020). *The case for degrowth.* Polity Press.

Pérez Orozco, A. (2014). Subversión feminista de la economía. Aportes para un debate sobre el conflicto capital-vida. *Cuadernos de Relaciones Laborales, 32*(2), 506–511. https://revistas.ucm.es/index.php/CRLA/article/view/46815/43927

Piepzna-Samarasinha, L. L. (2018). *Care work: Dreaming disability justice.* Arsenal Pulp Press.

Piepzna-Samarasinha, L. L. (2021). Still dreaming wild disability justice dreams at the end of the world. In A. Wong (Ed.), *Disability visibility* (pp. 250–261). Vintage.

Puig de La Bellacasa, M. (2017). *Matters of care: Speculative ethics in more than human worlds.* University of Minnesota Press.

Resurrección, B. P. (2017). From 'women, environment, and development' to feminist political ecology. In S. MacGregor (Ed.), *Routledge handbook of gender and environment* (pp. 71–85). Routledge.

Roelvink, G., & Gibson-Graham, J. K. (2009). A postcapitalist politics of dwelling: Ecological humanities and community economies in conversation. *Australian Humanities Review, 49*(May). http://australianhumanitiesreview.org/2009/05/01/a-postcapitalist-politics-of-dwelling-ecological-humanities-and-community-economies-in-conversation/

Saave-Harnack, A., Dengler, C., & Muraca, B. (2019). Feminisms and degrowth–alliance or foundational relation. *Global Dialogue, 9*(1), 29–30.

Salleh, A. (2010). From metabolic rift to 'metabolic value': Reflections on environmental sociology and the alternative globalization movement. *Organization & Environment, 23*(2), 205–219. https://doi.org/10.1177/1086026610372134

Sato, C., & Soto Alarcón, J. M. (2019). Toward a postcapitalist feminist political ecology approach to the commons and commoning. *International Journal of the Commons, 13*(1), 36–61. https://doi.org/10.18352/ijc.933

Schneider, F., & Pope, J. (2020, September 28). *Degrowth is not a passive critique but an active project of hope*. Uneven Earth. https://unevenearth.org/2020/09/degrowth/

Shiva, V. (1989). *Staying alive*. Zed Books.

Shukla, A., & Kulkarni, S. (2019). Harvest of the uteruses. *Economic & Political Weekly, 54*(29).

Simpson, L. B. (2017). *As we have always done: Indigenous freedom through radical resistance*. University of Minnesota Press.

Singh, N. M. (2019). Environmental justice, degrowth and post-capitalist futures. *Ecological Economics, 163*(September), 138–142. https://doi.org/10.1016/j.ecolecon.2019.05.014

Smith, T. S. J., Baranowski, M., & Schmid, B. (2021). Intentional degrowth and its unintended consequences: Uneven journeys towards post-growth transformations. *Ecological Economics, 190*, 107215. https://doi.org/10.1016/j.ecolecon.2021.107215

Tallbear, K. (2019). Caretaking relations, not American dreaming. *Kalfou, 6*(1), 24–41. https://doi.org/10.15367/kf.v6i1.228

Tuck, E. (2009). Suspending damage: A letter to communities. *Harvard Educational Review, 79*(3), 409–427. https://doi.org/10.17763/haer.79.3.n0016675661t3n15

Tuck, E., & Yang, K. W. (2012). Decolonization is not a metaphor. *Decolonization: Indigeneity, Education & Society, 1*(1), 1–40.

Whyte, K. P. (2020). Environmental justice, indigenous peoples, and consent. In M. Mascarenhas (Ed.), *Lessons in environmental justice: From civil rights to Black Lives Matter and Idle No More* (pp. 35–49). Sage.

Wichterich, C. (2014). Searching for socio-ecological and socio-economic transformation: A feminist perspective on the 4th De-growth conference in Leipzig. In *4th Degrowth Conference in Leipzig*. Leipzig. Retrieved May 31, 2022 from http://www.femme-global.de/fileadmin/user_upload/femme-global/themen/English/Care_degrowth_AEPF10.pdf

CHAPTER 9

Perspectives on Decoloniality for FPE

Dian Ekowati, Marlene Gómez, Iliana Monterroso, and Ankita Shrestha

Introduction

Our initial motivation to write this piece was to explore some of the core proposals of decolonial theories that can nurture feminist political ecology (FPE) theory and practice. But rather than engaging in a dialogue between FPE and decolonial thinking, which we recognise may be inexhaustible and therefore too vast for the scope of this chapter, we choose instead to piece together some of our intimate understandings of decolonial thought. The chapter is organised around four pieces that reflect aspects of the personal intellectual journey of the writer through their epistemic relationship with different experiences and understandings of coloniality, or, put simply, their reality as researchers.

The order of the authors is alphabetical order. It does not reflect the commitment and responsibilities of authors' contributions.

D. Ekowati
WEGO Innovative Training Network, University of Brighton, Brighton, UK
e-mail: Dee.ekowati@gmail.com

W. Harcourt et al. (eds.), *Contours of Feminist Political Ecology*, Gender, Development and Social Change,
https://doi.org/10.1007/978-3-031-20928-4_9

Our aim is to be transparent to the reader about the 'places we come from', both conceptual and literal, and therefore 'situate' ourselves in each of our reflections to show who is speaking from where. This chapter's layout also brings out the multiple voices and perspectives underlying the different colonial realities we all live as researchers from the 'global' South, navigating between research and activism. Each piece however discusses different issues. Through our effort to write a joint chapter for our collective, we acknowledged that our individual and unique trajectories have shaped how we understand the concepts of coloniality and how we subsequently attempt to decolonise our areas of research and ourselves. We think that our differences, which, instead of listing here, we seek to allow our readers to explore for themselves, are an outcome of thinking through decoloniality. Although the texts here may read as disconnected for those reasons, we do this exercise not just to bring forth the multiple interpretations of how decoloniality may be conceptualised, but also to highlight how both FPE and decolonial thinking can accommodate the many, and often contradictory, strands of non-extractive research approaches under the same roof.

Ankita Shrestha: On Coloniality and the Political Subject

To think 'the other' in a colonially drenched epistemic thought is to think *of* the other as an object (Mbembe, 2001). This 'otherness' is inflicted upon the native body in a colonial encounter. The coloniser appears with force, either with guns and artillery, or with knowledge that the native

M. Gómez (✉)
Freie Universität Berlin, Berlin, Germany
e-mail: marlene.gomez@fu-berlin.de

I. Monterroso
Center for International Forestry Research, Bogor, Indonesia
e-mail: I.Monterroso@cgiar.org

A. Shrestha
University of Oslo, Oslo, Norway
e-mail: anki.shrestha@gmail.com

must contend with. The coloniser does not ask for the native's permission; the coloniser simply violates. In this affliction, they force the native to accept their fate of becoming a *colonised, native* body—the other. In birthing the body of the 'native' into existence, the coloniser forges a common history in which he (*sic*) "owes the fact of his very existence to the colonial system" (Fanon, 2001, p. 28). Here, the nature of the encounter becomes almost irrelevant momentarily, as whether by force or with consent, neither the native nor the coloniser are the same again.

How then does one define coloniality[1] in a country like Nepal, one of the rare small nations that resisted all European colonial powers? The narrative of the 'post-colonial' nation has decidedly not been applicable to Nepal and has been reserved for neighbouring India and other smaller nations of South Asia that were actual colonies once. Yet, the ahistorical[2] technicality of being a sovereign nation sits oddly with the way developmentalist projects of modernity grips the imagination of an entire people and dictates its government. Complex relations of caste, class, religion, and gender lay bare, and often brutally so, the multiple dimensions of coloniality experienced in everyday social and political relations today. Below, I attempt to put forward my interpretation of who the 'decolonial' political subject could be, as I seek to avoid universalising all colonial experiences and to contribute to colonial scholarship that provides space for multiple temporalities to come forth.

The Decolonial Subject

Let us ask first who is the decolonial subject. Here, we are asking two ontologically motivated questions—*who* the subject is and *what* is the decolonial. The subject, in a most reductively poststructural sense, can be conceptualised as an individual, a body capable of engaging in power relations (Lukes, 2005) and of rejecting them (Butler, 2006). Outside a

[1] Aníbal Quijano's conceptualisation of 'modernity/coloniality' puts forward the idea of coloniality not merely as the state of being a colony but also refers to an epistemic form of colonisation through which Eurocentric ideologies such as modernity overpowers all other forms of epistemes.

[2] The colonisation of indigenous populations during the imperial campaign of the Shah rulers eventually led to the project of the modern Nepali nation-state (Regmi, 1999). These temporalities are often not seen as processes of colonisation because of the absence of its 'white European' coloniser at the centre.

phallocentric view of the subject, it is also a sexual body, whose difference against another body is characterised through the categories of male or female (Irigaray, 1995) or through historical constructs of gender (Mohanty, 1988, 2003) that not just inform struggles of identity and representation but also reveal the political economy of subsequent knowledge production (Asher, 2017). In a more existentialist view, which is of consequence here, the subject is a person, a sentient being, a body among bodies (Jung, 1996). But against dualistic views of the body as our medium *into* the world, or, put simply, our way of experiencing the world and of making sense of it, it is important to radicalise the body as not separate from but immanent to the world (Deleuze & Guattari, 2005) in order to think the 'decolonial'. For using the word decolonial today is to speak foremost of an attitude towards viewing the post-colonial world as a thinking people (Walsh & Mignolo, 2018), a people historically altered by the colonial encounter, but moving, for those exact reasons, towards a world which is not defined by (or reduced to) a colonial caricature, an epistemic residue of the coloniser, but by an ontological rebirthing of the native reality. In other words, it is an act—in thought and movement—that attempts to shed the colonial remnants of a brutal force that dominated, dehumanised, and decapitated the native body, to relive as the body.

If all three interpretations of the subject listed above could be distilled down to one, we could then say that by the very fact that it exists, *acts*, and through action (or the decision to not act), the subject occurs. We interpret the 'decolonial subject' then as follows. The decolonial subject acts to recuperate the body from coloniality (Quijano, 2000), an ontological imposition of the colonial encounter that irrevocably changed the social, economic, cultural, and, above all, epistemic systems of the colonised world such that Europe, or the geophysical spaces of the colonisers, and its ideals of modernity, became the cornerstone for all modes of civilisation that followed (Bhambra, 2014). The *decolonial body* then acknowledges colonisation as a historical process understood "in the exact measure that we can discern the movements which give it historical form and content" (Fanon, 2001, p. 27). Such movements must include, then, contingencies that skip the form and content that make up history, for these contingencies, although they are left outside the margins of history, are scarcely forgotten by the body. Indeed, everyday histories are resurrected in the experiential body; as such it does not experience coloniality but *is* that very experience, marked by temporal contingencies of colonial encounters that continue to occur in everyday life.

The Political Subject

If the task for the decolonising body is to recuperate the body, I contend that it needs the political. The 'political' is emphasised here as a body occurring as an ontological rather than an epistemic necessity. In the post-colonial world, this 'political' body is underscored by coloniality, through which colonial power regimes are not only reproduced in everyday human relations but encounters with the colonial past are also perpetually renewed in different relational spaces in the present (Mbembe, 2001). This 'political' body is therefore a body of *multiplicities*, unforgiving and unforgetting, because it refuses to be forgotten or forgiven itself, and yet also a body of *possibilities* capable of putting aside the burden that history has put on its shoulders. Without the political, the decolonising body, limited in its efforts to overturn the epistemic systems of power in society, culture, and the environment that persist under the paradigm of modernity, is restricted to the will to power, motivated either towards the obliteration or negation of the other (Sartre, 2007) or the transformation into the other (Merleau-Ponty, 1964). To reject assimilating the subject into the folds of coloniality/modernity is therefore not enough, as this cannot simply be achieved by reversing power orders or replacing power structures (Mbembe, 2001). I put forward then that the decolonial aim, instead of succumbing to the telos of power, could shift to transcending the limits of the decolonial body itself, as the political. This political body is motivated by the will to power which, however, is also the will to transcend itself (Nietzsche, 1967). I propose then that the decolonial subject extends itself towards the political, as it forces us to re-evaluate the categories of 'I' against 'the other' upon which the colonial encounter was legitimised; a political reimagination of these relations forces thus to think through the lens of forged realities, for the post-colonial world neither 'belongs to' the colonised or the coloniser but recreates itself every day through the body.

Valorising colonial encounters as a necessary step towards thinking of the other is not, however, the attempt of the decolonial thinking engaged here, and neither is the attempt being made to recognise colonial encounters as the only way to think of the other. I attempt here to facilitate an existential imagination of the decolonial subject such that we render the decolonial subject political. I use the word render not to suggest that the decolonial subject needs either to be reduced or refined to bring it to an altered, higher state of 'being political'. This would be tantamount to saying that other interpretations of the political, of the epistemic revolt and social movements of subversion of power that the decolonial world has been engaged in already for decades, are somehow inadequate or not worth pursuing. It is however to say rather defiantly that seeing the decolonial as political could help to bring this high pursuit of the decolonial project down to a more minuscule, everyday level of individuals capable of mutating into 'the political'.

Iliana Monterroso: On the Making of Political Subjects

This piece explores how these different understandings of the decolonial thought materialise in practice, and how feminist perspectives around the decolonial and political ecology can further our understanding of the political. The following section is based on personal engagement with research and development practice around forests. I was born and raised in the Guatemalan highlands in Totonicapan, an Indigenous region where forests are managed as *Parcialidades*.[3] Forests are not only the *focus* of my research, rather, but they also represent a sense of place and reference point that has strongly influenced my professional career. Indigenous forests in the highlands became the flagship of the K'iche' People,[4] a stronghold for Indigenous authority systems over forests, lands, and water resources. In this text, I argue that decolonial thought and practice emerge from and help understand the making of forests and those inhabiting forestlands as political subjects (Springate-Baginszki & Blaikie, 2013). I contend that struggles around the political forests and forest people are closely embedded in processes of constant negotiation of power and assertion of authority and legitimacy of both knowledge and practice. This entails the recognition of forests as relational places, inquiring about the boundaries of forest ecosystems as defined

[3] *Parcialidades* are forest systems managed as commons, with rights grounded on the negotiation of land titles since colonial times. These forests are governed by complex institutional arrangements based on kinship (Reddy, 2002).

[4] One of the Mayan ethnic groups recognised by the Guatemalan State.

by ecological and biophysical terms (Nightingale, 2018). Shaped also by the social norms, practices, and relations between humans and non-humans, the forest is itself embedded in power struggles and processes of territorialisation (Loivaranta, 2020).

Situated Engagement with Decolonial Thought and Forests

Colonial histories frame the way forests, and those women and men inhabiting forestlands, become political subjects. However, while some progress has been argued in forest studies, feminist decolonial and political ecology perspectives show the need to decolonise practices that continue commodifying forestlands and exacerbate exclusionary practices that reduce forests to commodities and forest people to environmentally responsible subjects (Agrawal & Bauer, 2005). This has implications both for framing our theoretical and methodological understanding of forests and for those living and depending on these resources. I start by laying out my journey into the studies of forests and those living in forestlands.

I pursued undergraduate studies in Biology as part of a US scholarship programme for Central American students. On my return, I joined a community forestry research programme. I travelled to the Guatemalan lowlands in the north, a frontier tropical forest region. I conducted community training in both the highlands and the lowlands. The contrast in forest landscapes in these two regions was as diverse as the social processes and community organisation I encountered. My understanding of forestlands as places for colonisation, management, and/or conservation shifted as I came to see them also as places of identity, history, and social relations. Forests became not only an issue of biophysical concern but rather highly contested and political spaces. Later, I travelled to Europe, to pursue postgraduate studies. I was drawn by critical perspectives of Latin American ecological economists and political ecologists like Leff, Escobar, and Alimonda who not only questioned mainstream perspectives but also provided situated knowledge to understand the complex histories (Escobar, 2006).

I kept relationships with local, forest, community, and Indigenous organisations, amazed by their diverse practices and histories. I worked with forest, environmental and biophysical scientists. Not all of them acknowledge the contextual elements of forests—the contested interests around them, the diverse knowledge local people had of them. In the field, until recently, I found very few women leaders, researchers, and practitioners. I was left with the feeling that research lagged far behind the processes of mobilisation and social change. My research work expanded to other countries that required engaging with other histories. This meant a constant process of unlearning practices of research, questioning the theoretical and methodological framings I used, and contrasting my own experiences (Asher, 2017).

The making of political forests and those of forest peoples provide examples of where and how feminist political perspectives can further avoid exclusionary narratives in the context of research around forests. Feminist decolonial and political ecology perspectives can better explain how forests can become a tool of power and what this implies for the recognition of social and political subjects. First, different territorialisation practices imply processes of negotiating authority and legitimacy that determines who can engage in these debates. Second, forest dwellers and their struggle to assert their recognition as political subjects, have long been framed by technical perspectives that have reduced what can be considered a forest, who can be considered a forest community, and for what purposes a forest is used. These reflections highlight how FPE perspectives can inform decolonial thinking, providing approaches to better understand processes of 'decolonization of the self' that go from 'becoming the subject' into being political.

The Making of Political Subjects in Forest Landscapes

The making of the political forests is closely linked to issues of legitimacy, what is legitimate and how legitimacy is constructed (Fraser, 2015; Habermas, 1975). Elements in contemporary governmental assemblages—"discourses, institutions, forms of expertise and social groups whose deficiencies need to be corrected"—are also evident in the extent to which these social groups have gained statutory formal recognition of their governance structures and their ability to engage with state and market actors (Li-Murray, 2014, p. 263). This framing of governance has been criticised for the instrumental and managerial approach, that focuses on prediction and lacks a proper understanding of the historical processes that shape power relations in which governance structures are embedded (Arts, 2014).

Differing from the Guatemalan highlands, the northern forests had long been considered hinterlands awaiting the modernising presence of the State. As in other tropical forests regions, commoditisation of forests, e.g., through the creation of markets for forest ecosystem services and other conditional incentive programmes to mitigate climate change, put forests back in the centre of national state and international policies (Sikor, 2010). However, very different approaches have been taken where options do not align with market and conservation interests. In the highlands and eastern forest regions of Guatemala, access and control of forests remain highly contested and often violently restrained (Ekern, 2006). In the north, conservation policies recognising forest communities as protected areas were established and different management models were discussed to recognise "environmental subjects". Forest communities' identities and knowledge systems were deemed interesting vehicles for implementing state forest policies (Agrawal & Bauer, 2005). This process of subject-making of forest-dependent groups shaped the establishment of protected areas, legitimising state authority in these forest hinterlands aligned with state conservation policies.

Power struggles underlying the contestation and negotiation process that followed the social mobilisation process often homogenised community groups, and diluted the diversity of age, gender, and ethnicity, reducing the ability to address underlying social differentiation and exclusion issues (Ybarra, 2017). In some regions, establishing arrangements between the state and the communities kept some communities from being evicted. In others, protected areas meant enclosures; green grabs that resulted in violent dispossessions with negative consequences for Indigenous communities, who were often portrayed as driving deforestation (Grandia, 2012; Ybarra, 2017). This shows how colonial legacies of structural racism strongly influence governmental environmental assemblages, creating new political subjects as needed and shaping them in such a way that maintains discriminatory practices and sustains extraction.

Divergent interests and power struggles have shaped which social groups are given the legitimacy to claim contested resources in forests. In the case of forest-dependent communities, the process was legitimised by the state and conservation NGOs based on their ability to meet conservation goals, often measured by deforestation rates, while at the same time this process consolidated protected areas as territorialisation policies. International policies that sought to consolidate state authority in forest frontier regions enforced these practices. This renewed interest in forests, as evident during COP 26, claims to recognise Indigenous peoples' role of "forest guardianship". However, engaging Indigenous peoples in these forest and climate change governance processes is closely tied to their recognition as political subjects. Recognising forest peoples as political subjects must entail decolonising existing technocratic approaches around forests, pushing the boundaries of both research and practice. As Mignolo would argue (2017), this requires "epistemic reconstitutions" to "change the structure of knowledge" challenged by decolonial epistemologies underlying the way we engage with forests.

Marlene Gomez: On Body, Territory, and Care

My first approach to decolonial theories began when I was 22 years old. Back then, I was engaged with environmental movements and peasant activism. I was also critical of the capitalist way of life—unconsciously consuming products in mass—and I tried to alter and reduce my consumption habits. I tried to live this way during my time as an undergraduate, until I came across Latin American decolonial theories that opened my comprehension of the real crisis we were living in, which is not only institutional but civilisational. Reading Quijano, Lander, Walsh, Acosta and other decolonial thinkers informed me that the colonisation of Abya Yala[5] (America) was only possible through the imposition of a modern rationality that comes along with the hierarchisation of societies based on racism, command–obedience relationships, gender differences, and the exploitation of nature. This modern rationality is Eurocentric, capitalist, and patriarchal and organises the modern world through the validity of epistemologies and ontologies developed in Europe. Back then, approaching decolonial theory upended my project to save the planet by contesting State politics and made me question privileges, positionalities, and possibilities, turning towards how we could care for others and the planet along with a process of decolonisation.

In this piece, I want to focus on the proposals of decolonial feminism in three different strands: the body, territory-territoriality, and care for and with the other. These proposals show that caring for the planet and others are related to the ways we engage with the world in our everyday life. Firstly, I will draw on the body as a scale of collective resistance, a host of diversity, and a vehicle for caring for and with others. Secondly, I will explore notions of territoriality that show relational and plural understandings of life. Thirdly, I will briefly touch upon political projects that

[5] As this territory was named by the Second Continental Summit of Indigenous Peoples and Nationalities in 2004.

decolonial feminism envisions as caring pluriverses. This way, I will try to present narrative others that aim to dismantle care as a pure class-based or economic relation and position it as a powerful vehicle for social transformation. By this, I do not mean to neglect what feminist economists eloquently inform us about care and how it changes along lines of gender and race, producing labour inequalities. However, I want to draw on narratives that imagine care as work decentred from economic dynamics and out of the realm of *homo economicus*. Clues for this can be found not only in decolonial feminism, but also in feminist political ecologies, post-capitalist approaches, the commons, and other approaches that look at the roots of communitarian bonds to rebuild the social fabric based on careful collective values among people and the environment.

Situating Decolonial Feminism

Dismantling the patterns of power that constitute the modern world is the challenge of decolonial feminism in Latin America or Abya Yala. Decolonial feminism is not only a social movement, but also "a symbolic space of cultural affirmation, identity formation, knowledge production, and social and political action" (López Nájera, 2014, p. 108). Questioning patterns of power such as race, sex, gender, class, and ethnicity, which hierarchically organise societies, decolonial feminism shows the multiple facets that systems of oppression exert in our bodies, territories, and everyday life (Espinosa-Miñoso, 2014). Taking up the knowledge, practices, and organisational forms of original peoples, decolonial feminism seeks to give a pluricultural, epistemological, and ontological recognition of the diverse identities, struggles, and resistances that make up the territory of Abya Yala (Cumes, 2009). Under threat from oppressive or colonial power structures, listening to common sense, recognising knowledge production in daily practices, and embracing plural dialogues and imaginaries, are the vehicles and tools that nourish a political project for change. Nowadays, the constant struggle against femicides, sexual and gender violence, inequalities, and the disciplining of bodies unite the cuir, trans, lesbian, anti-colonial, and decolonial feminisms of the region.

The Body

Decolonial feminism explores how violence is inscribed in bodies as a colonial legacy. Decolonial feminists work to make visible bodies fleeing from the heterosexual regime, gender inequalities, white hierarchies, body consumption, and coloniality (Espinosa-Miñoso, 2014) in a world where the body is a container of systems of oppression and hierarchies, besieged by multiple violences. Understanding the modern body as a colonial legacy allows us to understand it in relation to a territory which was also colonised. The body, from a decolonial and feminist decolonial perspective, is a microscale of the colonised territory of Abya Yala and globalised hierarchies of race and phenotype (Quijano, 2009; Segato, 2004). The global imposition of modernity introduced a binary understanding of life within societies ruled under the complementary of dualities (Lander, 1997; Segato, 2004). This reconfigured the relationships between subjects and nature, bringing the Cartesian model as the legitimate method to understand and engage with life. Through the body, it is possible to inhabit the colonisation of lands and nature that provokes displacements and the introduction of external ways of life that create conflict and violence.

However, with the body we take the streets, we defend lands from mining, we resist the Other, and we practise care for and with others. This is possible because the body is built through reciprocal practices among humans and material and immaterial components of life that are orally and corporally defended. We can see this in intergenerational practices that protect collective histories and in social imaginaries that inhabit and practice ways of life that resist subsumption to colonisation, such as cultural identities and political projects like the *sentipensar*, *Buen Vivir*, or *Sumak Kawsay*. This is not far from thinking that the body is permeable (Marcos, 2008) meaning it is open; created and recreated by a constant contagion and complementarity with other bodies. Following this, we can assert that the body is in constant historical and political production, through which it can resist and become one of the material vehicles for social transformation. Through our body we show our story and negotiate how we inhabit space and resist invisibilisations. Through feelings, discomfort, and hierarchies, the body is intersected by patterns of power that are historically related to our territory; patterns which trigger a systematic increase of violence. This informs how the history of communities is inscribed in the bodies of the people who inhabit them: we see this in the thick hands of a care worker washing other people's clothes, in the rough hands of a peasant, or in the hands of a woman who daily carries water in long distances. Through the expansion of colonisation, the body continuously becomes a territory of conquest, but also a territory of resistance, where patterns of power and systems of oppression are daily inhabited, perceived, and experienced but also negotiated through skin colour, daily habits, diets, language/accents, among others.

Territory-Territoriality

The analysis of the territory in decolonial theories questions the traditional understanding of territory as a place delimited by borders that contains a fixed and neutral nation-state. Critiques of this understanding argue that the nation-state is a political force based on modern practices that homogenise subjects and territories, masked as citizenship and acculturation (Yuval-Davis, 1993). Decolonial feminism is situated in this critical strand and, influenced by Indigenous, peasant, and other diverse movements, considers the territory to be a living place that obtains meaning through the bodies that wander, defend, and resist with it. For this notion, the territory is a tool of power and not a fixed container of it, regulated only by institutional norms and legal frameworks. To the contrary, the territory is a place of contingencies, contradictions, and antagonisms that allow for transformations, joy, and convivialities that construct the meaning of the nation-state. Although the territory is the space that can be conquered, it is also the place that witnesses resistance and negotiation. This is possible because the territory is part of the social fabric, along with the rest of the components of life—water, plants, and everything that composes the socio-natural network. For such a conception of the territory, the notion of community plays an important role, meaning that life is built in a co-constitutional and relational way among humans and among humans and nature.

By thinking of the territory as a living space, we must acknowledge its capacity for mobility, a characteristic called territoriality. The experiences of communality, a political project of peasants and Indigenous peoples in Oaxaca, Mexico demonstrates this notion: these communities understand territoriality by the four pillars that compose the community: the assembly, the territory, the festivities, and the *tequio* or community work (Luna, 2003), practices that are inhabited by the bodies and social fabric of the community members. What gives the idea of mobility to territoriality is the fact that these practices can be brought along wherever the community members go by their orality, behaviour, imaginaries, and ways to relate to the territory. In this comprehension, territoriality is never neutral; by its mobility, it is intersected by patterns of power, systems of oppression, and violence that bodies experience in conditions of race, class, gender, ethnicity, among others. Thus, the body and the territory are co-constitutional of a reciprocal life and are travellers through our territoriality; so is the way we relate to the place where we are.

Caring for and with

Within decolonial feminism, the debate on care work is a central axis composed of different strands. Some understand care from a feminist economic perspective and highlight the role of productive and reproductive work within a patriarchal framework (Díaz, 2009; Henrich, 2016). The dialogue with Black feminism nurtures decolonial feminism through intersectionality, emphasising patterns of power and colonial legacies that keep care work dynamics unequal. Decolonial feminism notes that care needs to be critically rethought from a communitarian perspective (Millán, 2019) and through using critical pedagogy (Walsh, 2015) where the environment plays a central role. In this debate, domestic economic units, solidarity economy, reciprocity, and *Sumak Kawsay/Suma Qamaña*, and *Vivir Bien* practices, question the relationships, dynamics, and ethics of care through which we relate as humans and to the components of life. Here, the subject is one that feels and that claims the right to live in dignity among other subjects and nature in a relational manner. However, these practices have undergone a strong critique from decolonial feminism since they are in most cases subsistence work carried out through unequal dynamics and by diverse subjects. Here, just as in households, gender identities and working roles are constructed, reinforced, and configured that make care relations unequal and problematic. The proposal for this is to reassess and rearticulate care dynamics from a collective perspective where care is not unidirectional but multidirectional (Gómez-Becerra & Muneri-Wangari, 2021) in connection to body and territory needs.

Situating struggles historically and geographically allows us to illuminate how care practices are negotiated through the politics of everyday life, the body, and the territory. Decolonial feminists make the call to think about care relations from reciprocity and solidarity through which we can consolidate collective caring practices. Questioning who is able to enter into the dynamics of care, how care is exercised, and why it is needed must be a daily practice, wherein a collective manner of relating to subjects and nature sets the pace for these relationships. Care cannot be seen anymore as a human-centred relation, but as an expanded relational practice nurtured and contested by diverse bodies, territories, and territorialities. Once we collectivise care practices, they can act as one of the vehicles to decolonise our everyday experience of colonial legacies, immersed in modern, Eurocentrist, capitalist, and patriarchal relations, bodies, territories, and institutions.

Dian Ekowati: On Care in the Indonesian Oil Palm Community

My Understanding of Decoloniality

I was first exposed to the idea of decoloniality when I started my PhD in feminist political ecology in 2019. I remembered that one of the first articles I read on decolonial ideas was from Chandra Talpade Mohanty: in her article *Under Western Eyes* (Mohanty, 1988), she eloquently argues against the monolithic categorisation of "Third World Women" and the othering, colonial assumption behind this categorisation. Later on, when I read Edward Said's work of post-colonial theory, "Introduction to Orientalism" (Said, 1978), and Linda Tuhiwai Smith "Decolonising Methodologies" (Smith, 2012), I was exposed to similar conceptions of "otherness".

Throughout my Ph.D. I read works from other scholars who also work on the Indonesian oil palm industry, had a similar organisational background in development and who shared the spirit of decoloniality. I found this excerpt to sum up what I feel at this stage of my life and career/study as a fourth-year Ph.D. student about being a decolonial feminist researcher.

> The commitment to change and attention to the relationships with research subjects are key to feminist research. Practices of **reflexivity** – **researchers** reflecting on their **positionality**, **critically examining the research process**, and the **commitment to change** (Hesse-Biber, 2014), and attention to the **relationships between the researcher and**

> **the research subjects** (Stacey, 1988; Nagar, 2003; Craven & Davis, 2013)—stand out as the defining characteristics of feminist research. (Resurrección & Elmhirst, 2020, p. 151, emphasis added)

I use the excerpt as a reminder of how significant positionality and reflexivity are in looking at the research process and commitment to change, something that I find resonant with the commitments behind decolonial ideas; despite the fact that none of the above authors mention the term "decolonial" in their chapter.

Below is my attempt to be reflexive towards my own positionality as an attempt to decolonise my research on the topic of care in the Indonesian oil palm community. I explain this by looking back and looking forward at my life stage, work, and research journey, hopefully to ease the reader in understanding which is my reflective and reflexive attempt (looking back), and which is my decolonial attempt (looking forward).

Looking Back: Reflecting on My Positionality

My positionality affects the way I make sense of the topic of my research on care, and engaging reflectively with my positionality and realising the power relations between researcher and subjects are the first steps towards thinking decolonially.

I remember proposing beyond human care at the start of my Ph.D. Later on, when I was in Indonesia, I decided to change my focus to care in everyday life, as I found the meaning of care in the oil palm context as imagined by Global North consumers did not travel well to the actual communities of oil palm producers in the Global South. My position as an "insider" to the experience of everyday care in the communities I worked with drove me to that change. I was born and raised in a small village without electricity until I was 11 years old. This experience is essential in my research practice since not having access to electricity is often the most notable aspect mentioned by local communities when they speak about the feeling of being "backward" or in the dark (Elmhirst et al., 2017). My parent's home did not have household appliances (e.g., fridge, washing machine, cooker) until I was 23 years old. I helped my mother (my father did little care work for his children) to care for my two younger siblings (with a three- and twelve years age gap). I find the experience of caring for young children and not having access to appliances is essential in my understanding of care. Domestic work is mundane, exhausting, and mainly undervalued; despite it being vital to care for and maintain life.

Later on, I got married and had two young children. Today, I have both the experiences of living every day with care support and the experience of lacking support, due to COVID, finances, and life circumstances. During COVID, my husband and I worked full-time from home while taking care of our 1.5- and 7-year-old children. Schools and childcare were closed, and we could not ask our caretaker to come to our place in order to avoid risks (for our caretaker and for us). It was hard, physically and emotionally, especially since the four of us were living in our two-bedroom house with little outdoor space.

This embodied experience as a daughter, sibling, wife, and mother with and without care support, shaped my understanding as "insider" to my own research on everyday care realities and challenges.

Throughout my life, I have had the privilege to travel and learn from agricultural, forest, and oil palm communities, mostly in Indonesia. My position is not always the same and my relationship with oil palm communities has been different from what I have now, i.e., as a Ph.D. researcher trying to understand various oil palm community's care discourses using online videos/documents. I had been working as a junior researcher and research assistant in national and international universities and NGOs for almost 15 years before starting my Ph.D. I did fieldwork and during my work planning[6] I often had an awkward mixed feeling of desire to help the community while recognising that our institution's worldview and methods were undermining the community's worldview. Sometimes this was not something we intended, but pressure from the donor and project budget and timelines often left us no room to think or do otherwise. Later, when we would visit the community, there was often another awkwardness—the community called us all from "Jakarta '' or the "centre", while they are in the "margin", they are "the other"; implying that we are different, worlds apart. Jakarta is Indonesia's capital, where power, information, and facilities are centred. Carol J. Pierce Colfer, a researcher focused on gender and development, describes a similar feeling of awkwardness and power differentials in the context of her work with forest communities (Resurrección & Elmhirst, 2020).

[6] Where I mostly did not have a voice in the work objectives and approach—I saw myself as in the margin due to my Indonesian education/work background, as a young woman and in my position as assistant/junior.

Other than this, I am of the Javanese ethnicity—Javanese are seen as the "whites" of Indonesia, in the sense of occupying the majority and living where power is centred, but also as one ethnicity that is often seen as most patriarchal in Indonesia. On top of that, I am a Moslem (the majority faith in Indonesia). There are long colonial historical instances of oppression and discrimination from Javanese ethnicity (the centre) towards **other**[7] ethnicities and Moslem (the centre) towards **other**[8] faiths.

All these reflections on positionality and my journey with decoloniality remind me of the terrible feeling about how our team did our work before. How we saw the "other" in the communities we worked with, how the "others" were different from us, and how sometimes I witnessed that our team leader did something that they would not do to us, but our leader did to the community anyway. This is something that I am not proud of remembering up to this day. I remembered in one instance, my senior lecturer who led the community visit was welcomed by an Indigenous leader on his porch, where they sat for two hours. When we left, this lecturer said "See, I can already make five journal papers about this *adat* (Indigenous) community from my two-hours talk with him". I remembered being so puzzled by his comments, although I did not have the vocabulary of 'extractive' and 'colonised' research to describe them back then.

My position as both an "insider" and a researcher exposed me to other forms of coloniality in research: in 2021, several researchers from the Global North expressed their frustration at not being able to enter Indonesia for research (one asked me personally about possible strategies to enter). This was at a time when COVID was at its peak, and deaths in Indonesia were exorbitant. I was puzzled by these questions, on how these researchers blatantly ignore the health risk they would bring to the community.

[7] Other is in bold to show that all other ethnicities other than Java in Indonesian is "the other" in the sense of "othering" as in post-colonial and decolonial works.

[8] Other is in bold to show that all other faith other than Islam in Indonesian is "the other" in the sense of "othering" as in post-colonial and decolonial works.

Looking Forward: Situating Positionality and Reflexivity in Decolonialising Research

I ask myself; how do I perform decolonial research? Which, in line with my journey above, I translate to "How does one do careful research without extracting and colonising the community, bringing marginal voices to the centre; all during the pandemic time?".[9] In my fieldwork, this entails visiting remotely located oil palm communities, and in Indonesia, it means very little to no access to health facilities.

In a different context, it reminds me of what Linda Tuhiwai Smith mentioned in her book about how colonisers (who judged their mission as much more important than Indigenous people's lives) sent infected blankets to Indigenous peoples in the First Nations of Canada (2012, p. 65). Therefore, I decided not to judge my research as so crucial that risking others' life is acceptable. This ethical stance led me to decide not to do my research face to face even when travel restrictions were relaxed by the Indonesian Government.

With this realisation, I decided on a YouTube ethnography, where I follow some young women oil palm farmers. They regularly post videos on their public channels (with millions of subscribers and hundreds of thousands of views of each video they post) about their everyday life as oil palm farmers in different places in Indonesia. I made a deliberate choice to follow young women who make public YouTube videos and who identify themselves as oil palm farmers. I made these choices as my research is about care and women traditionally carry more care burdens.

In deciding on everyday YouTube ethnography, I made a deliberate decision to only choose the channels that were intentionally setting their YouTube for wide audiences/public settings and to focus on voices that rarely make their way to centre stage in oil palm discourse: those of women and youth. Furthermore, in Indonesia, local YouTube channels are mediums that capture a lot of audiences across Indonesia. Indonesian mobile internet service provides many affordable YouTube packages for its users. However, this voice rarely enters mainstream discourse, either in academics or policy.

[9] Questions surrounding non-extractive research during pandemic times was also asked by Dupuis in her research with aged people (in Harcourt et al., 2022).

On the flipside, I ask about the ethics of "lurking" these spaces and the border between public and personal space of these persons I watch/observe (Morrow et al., 2015) something that I continue to reflect on now when I write this piece.

Conclusion

Through this chapter, we intend to contribute to the growing South–South dialogues which contest colonial and neo-colonial social and political structures in order to create a shared decolonial future. In our different interpretations of this decolonial future, we see a common spirit that links us all to FPE's attempts to grapple with complex questions of who the subject can be, and how to think about care for a common world. Our aim is to expose our different interpretations as a necessary step to engaging different ways not just of thinking about and engaging with decoloniality, but also of articulating these interpretations. These interpretations are bound within our understanding of the concepts of subjectivity, the body, and the other. Our diverse understandings bring our unique research approaches and our visions of decoloniality, which is why each author speaks in the 'I' of the first person. This 'I' is also a unique reflection of our *shared* understanding of situatedness, that we understand collectively as our historical and temporal rootedness, and our shared views on power inequalities that were not just inherent to the colonial world but are fundamental to an ongoing struggle in a post-colonial world. Our South–South 'dialogue', then, is as much a dialogue within ourselves—our experiences, personal histories, and reflections of our individual colonial realities—as it is between our collective interpretations of them. Our goal is then to be vigilant about our own situated realities and meaning-making processes, as ethical considerations of care engendered by those meanings and shared subjectivities have shaped our past and continue to shape our futures.

Funding: This chapter was funded by the Wellbeing Ecology Gender and cOmmunities Innovation Training Network (WEGO-ITN) funded by the European Union's Horizon 2020 research and innovation programme under the Marie Sklodowska-Curie grant agreement No. 764908-WEGO 2018-2021.

References

Agrawal, A., & Bauer, J. (2005). Environmentality: Technologies of government and the making of subjects. *Ethics and International Affairs, 19*(3), 116–118.

Arts, B. (2014). Assessing forest governance from a 'Triple G' perspective: Government, governance, governmentality. *Forest Policy and Economics, 49*, 17–22.

Asher, K. (2017). Spivak and Rivera Cusicanqui on the dilemmas of representation in postcolonial and decolonial feminisms. *Feminist Studies, 43*(3), 512–524.

Bhambra, G. K. (2014). Postcolonial and decolonial dialogues. *Postcolonial Studies, 17*(2), 115–121.

Butler, J. (2006). *Gender trouble: Feminism and the subversion of identity*. Routledge.

Cumes, A. (2009). "Sufrimos vergüenza": Mujeres k'iche' frente a la justicia comunitaria en Guatemala. *Desacatos, 31*, 99–114.

Deleuze, G., & Guattari, F. (2005). *Qu'est-ce que la philosophie?* Minuit.

Díaz, N. Q. (2009). Economías feminista, social y solidaria: respuestas heterodoxas a la crisis de reproducción en América Latina. *Íconos: Revista de Ciencias Sociales, 33*, 77–89.

Ekern, S. (2006). Saving the forest through human rights: Indigenous rights and ethnic tension in Guatemala. *International Journal on Minority and Group Rights, 13*(2–3), 171–186.

Elmhirst, R., Siscawati, M., Basnett, B. S., & Ekowati, D. (2017). Gender and generation in engagements with oil palm in East Kalimantan, Indonesia: Insights from feminist political ecology. *Journal of Peasant Studies, 44*(6), 1137–1159. https://doi.org/10.1080/03066150.2017.1337002

Escobar, A. (2006). Difference and conflict in the struggle over natural resources: A political ecology framework. *Development, 49*(3), 6–13.

Espinosa-Miñoso, Y. (2014). Una crítica descolonial a la epistemología feminista crítica. *El Cotidiano, 184*, 7–12.

Fanon, F. (2001). *The wretched of the earth*. Penguin Books.

Fraser, N. (2015). Legitimation crisis? On the political contradictions of financialized capitalism. *Critical Historical Studies, 2*(2), 157–189.

Gómez-Becerra, M., & Muneri-Wangari, E. (2021). Practices of care in times of COVID-19. *Frontiers in Human Dynamics, 3*(34). https://doi.org/10.3389/fhumd.2021.648464

Grandia, L. (2012). *Enclosed: Conservation, cattle, and commerce among the Q'eqchi'Maya lowlanders*. University of Washington Press.

Habermas, J. (1975). *Legitimation crisis*. Beacon Press.

Dupuis, C. (2022). Feminist ethics amid Covid-19: Unpacking assumptions and reflections on risk in research constance. In W. Harcourt, K. van den Berg, C.

Dupuis, & J. Gaybor (Eds.), *Feminist methodologies experiments, collaborations and reflections*. Palgrave Macmillan.
Henrich, I. F. (2016). Economía feminista y economía solidaria: Alternativa al patriarcado. In J. L. Coraggio, J.-L. Laville, I. Hillenkamp, I. Farah, J. Jiménez, S. Vega, L. Guridi, & J. C. Pérez de Mendiguren (Eds.), *Economía social y solidaria: Conceptos, prácticas y políticas públicas* (pp. 83–105). Hegoa.
Irigaray, L. (1995). The question of the other. In L. Huffer (Ed.), *Another look, another woman* (pp. 7–19). Yale French Studies.
Jung, H. (1996). Phenomenology and body politics. *Body & Society, 2*(2), 1–22.
Lander, E. (1997). Modernidad, colonialidad y postmodernidad. *Estudios Latinoamericanos, 8*, 31–46.
Li-Murray, T. (2014). Fixing non-market subjects: Governing land and population in the global south. *Foucault Studies, 18*, 34–48. https://doi.org/10.22439/fs.v0i18.4650
Loivaranta, T. (2020). Post-human lawscapes of Indigenous community forests in Central India. *The Geographical Journal, 186*(3), 288–299.
López Nájera, V. R. (2014). Feminismos y descolonización epistémica: nuevos sujetos y conceptos de reflexión en la era global. In *Más allá del feminismo: caminos para andar* (pp. 99–118). Red de Feminismos Descoloniales.
Lukes, S. (2005). *Power: A radical view* (2nd ed.). Palgrave Macmillan.
Luna, J. M. (2003). *Comunalidad y desarrollo*. CONACULTA.
Marcos, S. (2008). *Religión y género*. Trota.
Mbembe, A. (2001). *On the postcolony*. University of California Press.
Merleau-Ponty, M. (1964). *Le visible et l'invisible: Suivi de notes de travail*. Gallimard.
Millán, M. (2019). Genealogías descolonizantes en los feminismos emergentes: el impacto de las mujeres zapatistas. *Revista Punto Género, 11*, 18–33.
Mignolo, W. D. (2017). Coloniality is far from over, and so must be decoloniality. *Afterall: A Journal of Art, Context and Enquiry, 43*(1), 38–45.
Mohanty, C. (1988). Under western eyes: Feminist scholarship and colonial discourses. *Feminist Review, 30*, 61–88.
Mohanty, C. (2003). "Under western eyes" revisited: Feminist solidarity through anticapitalist struggles. *Signs: Journal of Women in Culture and Society, 28*(2), 499–535.
Morrow, O., Hawkins, R., & Kern, L. (2015). Feminist research in online spaces. *Gender, Place and Culture, 22*(4), 526–543. https://doi.org/10.1080/0966369X.2013.879108
Nietzsche, F. (1967). *On the genealogy of morals*. Random House.
Nightingale, A. J. (2018). The socioenvironmental state: Political authority, subjects, and transformative socionatural change in an uncertain world. *Environment and Planning E: Nature and Space, 1*(4), 688–711.

Quijano, A. (2000). Coloniality of power, eurocentrism, and Latin America. *Nepantla: Views from South, 1*(3), 533–580.

Quijano, A. (2009). Des/colonialidad del poder: el horizonte alternativo. *Estudios Latinoamericanos, 25*, 27–30.

Reddy, S. P. (2002). Communal forests, political spaces: Territorial competition between common property institutions and the state in Guatemala. *Space and Polity, 6*(3), 271–287.

Regmi, M. C. (1999). *Imperial Gorkha*. Adroit Publishers.

Resurrección, B. P., & Elmhirst, R. (2020). Negotiating gender expertise in environment and development: Voices from feminist political ecology. In B. P. Resurrección & R. Elmhirst (Eds.), *Routledge studies in gender and environment*. Taylor & Francis.

Said, E. (1978). *Orientalism*. Pantheon Books.

Sartre, J. P. (2007). *Existentialism is a humanism*. Yale University Press.

Segato, R. L. (2004). Territorio, soberanía y crímenes de segundo Estado: la escritura en el cuerpo de las mujeres asesinadas en Ciudad Juárez. *Debate Feminista, 37*(April 2008), 78–102.

Sikor, T. (2010). Forest justice: Towards a new agenda for research and practice? *Journal of Integrative Environmental Sciences, 7*(4), 245–250.

Smith, L. T. (2012). Decolonizing methodologies: Research and Indigenous peoples. *Zed Books Ltd*. https://doi.org/10.4324/9781315700779-29

Springate-Baginski, O., & Blaikie, P. (Eds.). (2013). *Forests people and power: The political ecology of reform in South Asia*. Routledge.

Walsh, C. E. (2015). Decolonial pedagogies walking and asking. Notes to Paulo Freire from AbyaYala. *International Journal of Lifelong Education, 34*(1), 9–21.

Walsh, C. E., & Mignolo, W. D. (2018). *On decoloniality: Concepts, analytics, praxis*. Duke University Press.

Ybarra, M. (2017). *Green wars: Conservation and decolonization in the Maya forest*. University of California Press.

Yuval-Davis, N. (1993). Gender and nation. *Ethnic and Racial Studies, 16*(4), 621–632.

CHAPTER 10

Debating Population in and Beyond Feminist Political Ecology

Mila Fenner and Wendy Harcourt

Introduction

Population is rarely discussed in feminist political ecology. It is tiptoed around, perhaps because of all the emotions it evokes. In order to open up this debate, in this chapter we look at the heated responses to the renowned feminist Science and Technology Studies (STS) scholar, Donna Haraway's call to 'Make Kin Not Babies' (2015). Disagreements in academic debates are often motors for new knowledge and understanding (Collins, 2000, 2002). However, such disagreements can become bogged down in disciplinary dogma and semi-interpersonal conflict (Barney, 1990; Morgan & Baert, 2015). Constructive dialogue stalls, dismissive attitudes grow and certain opinions are relegated to the sphere of taboo

M. Fenner · W. Harcourt (✉)
International Institute of Social Studies of Erasmus,
University Rotterdam, The Hague, The Netherlands
e-mail: harcourt@iss.nl

M. Fenner
e-mail: fenger@iss.nl

W. Harcourt et al. (eds.), *Contours of Feminist Political Ecology*,
Gender, Development and Social Change,
https://doi.org/10.1007/978-3-031-20928-4_10

(Collins, 2002). When conflicting discourses confront each other after years of effective silence, emotions explode. Donna Haraway experienced such explosions firsthand when she re-introduced to (anti-racist, anti-colonial) feminism the proposal that future human population growth could be detrimental to human and more-than-human life (Clarke & Haraway, 2018; Haraway, 2015, 2016). For the sake of survival and 'multispecies reproductive justice', Haraway argued that she would like to see human numbers wind down to around 2 or 3 billion people through a voluntary reduction in birth rates, especially among the rich (Haraway, 2016, p. 103). Her attempts to bring population concerns back into feminist discussions were received badly. She was accused of taking "a decisive turn towards a primitivism-tinged, misanthropic populationism" and "trafficking irresponsibly in racist narratives" (Lewis, 2017).

Since Malthus, debates about the impact of human population size on the environment and on the viability of poverty alleviation have held widely opposing views (Bashford, 2014). In the last few decades, feminist academics have been at the forefront of exposing misogynistic, racist, and neo-colonial underpinnings of Malthusian thinking and some of the population programme policies aimed at the global South (Hartmann, 1998; Ojeda et al., 2019; Sasser, 2018). As a consequence of this genealogy of population critique, the possibility that there could be negative environmental and humanitarian impacts of future increase in population has been not on the table in feminist thinking on population.

These views are reflected in international policy circles (Campbell, 2007; Coole, 2013; Halfon, 2006). Due to efforts of the international women's movement, among others, at the landmark international UN Conference on Population and Development held in Cairo in 1994, international policy focus moved away from a discourse of the "population bomb" or "overpopulation" towards the concept of sexual and reproductive health and rights (SRHR) (Campbell, 2007). Now, with the increased concern about climate change, environmental degradation, international migration and growing global poverty and inequality, the population question is making its return (Coole, 2018).

In this chapter, we explore the explosive responses of feminists to Donna Haraway's thesis of "making kin not babies". We pay attention to emotions and how they play out in intellectual debates, and in population debates among feminist thinkers specifically. We do this in an informal dialogue format, where we engage each other on both a personal and scholarly level in order to map out the contours of feminist thinking about the fraught topic of population growth and population control. As we explain below, we write from different positionalities and with contrasting views on population policy. We have tried to meet each other right at the fault line between our views in order to explore our own emotions around the population question.

As Kathy Charmaz notes about feminist research writing: 'Increasingly, we appear in our texts as thinking, acting—and feeling—participants rather than as disembodied reporters of collected facts. Lines between the subjective and objective blur' (Charmaz, 2012, p. 476). Inspired by this blurring—of subjective and objective, of researcher and researched—we experiment with a semi-informal dialogic writing style to illustrate the experiential and embodied nature of doing FPE research. We aim to show in the chapter how the feminist research process is one of continuous learning and unlearning.

We have tried to forge a mutual understanding by cutting through academic disciplinary boundaries and the inevitable use of disciplinary jargon. Our focal questions are: why is it so hard to engage in dialogue on population? Has Haraway's recent call to 'make kin not babies' helped to change this? We are interested to see how emotions can shut down dialogue and how by paying attention to this we can open up rather than close debates around population.

We first introduce in more depth how FPE relates to Donna Haraway's slogan of 'Make Kin Not Babies'. We then write about the ways in which our different positionalities are linked to the divergent views we hold, before beginning the centrepiece of this chapter: our conversation focused on conflict and emotions in population debates. We decided to structure the dialogue under the headings of five 'primary emotions': surprise, anger, fear, sadness and joy, in order to connect our personal experiences with the different debates in feminist thinking around population.

Feminist Political Ecology and the Call to 'Make Kin Not Babies'

FPE scholarship raises questions about methods and ethics that relate to embodiedness, emotions, subjectivities and intimacy when seeking to understand environmental questions. Such attention helps us to deal with difficult questions that must be asked when we wish to engage in transformative research and to create more equitable worlds (Elmhirst, 2011; Hawkins et al., 2011; Mollett & Faria, 2013; Sundberg, 2004).

FPE "recognizes the interconnectedness of all life and the relevance of power relations–including gender relations–in decision-making about the environment" (Rocheleau et al., 1996, p. 296). As such, FPE scholars have a particular interest in epistemological tensions and embrace multiple methods and activist work (Harcourt et al., 2022; Richardson-Ngwenya & Nightingale, 2018). As a field critically concerned with gender, the environment and reproduction, FPE has engaged in issues around population (see for example the work of Mehta et al., 2019; Sasser, 2018). With its focus on emotion, subjectivity and intimacy, an FPE approach helps us to go below the big picture questions to how population is shaped on smaller intimate scales. The attention to emotions allows us to look at how people engaged in debates around population relate to each other and what epistemological consequences this has had.

Donna Haraway blew new controversy into the population debate in 2015. As we stated above, the provocations of Haraway and others, including Adele Clark (Clarke & Haraway, 2018), were met with resistance. In the clash among feminist and environmentalist discourses on population, Haraway's intervention is important. Haraway challenges feminists "to make 'kin' mean something other/more than entities tied by ancestry or genealogy" (Haraway, 2015, p. 161) and to "find ways to celebrate low birth rates and personal, intimate decisions to make flourishing and generous lives (including innovating enduring kin—kinnovating) without making more babies urgently and especially, but not only, in wealthy high-consumption and misery-exporting regions, nations, communities, families, and social classes" (Haraway, 2015, p. 164).

What we are also interested in here is how the responses among feminists to this work, both oral and written, have been rife with emotion. The affection and love many feel for Haraway and her work has informed the controversy (Hamilton, 2017; Lewis, 2017; Schultz, 2021; Torracinta, 2017; Turner, 2017). And the strength of the response demonstrates the hold of anti-population policy sentiment by many feminists engaged in debates on population.

Positioning Ourselves in the Dialogue

Before we dive further into our dialogue about the emotions that have surfaced around the slogan 'make kin not babies', we first need to position ourselves more transparently, especially as we are using our own emotional responses as part of the dialogue. Wendy has worked as an activist scholar with decades of engagement in feminist debates around the colonial, racist legacy of populationist thought. In her writing and advocacy for sexual health and reproductive rights issues she has expressed deep concern about the instrumentalisation of women of colour in discussions of population numbers, poverty and environmental degradation (Harcourt, 2009, 2020). Milla is an early career scholar currently conducting her Ph.D., is trained in ecology, conservation and demography and is closer to academic discourses that speak of growing numbers of humans as a likely future contributor to extreme poverty in certain places and, to a more limited extent, a potential strain on specific environments.

We have been working together as supervisor and Ph.D. student since 2018, though over these years together we have found ways to go beyond that specific hierarchical relationship. We engage as two women of different generations sharing concerns about our health and well-being, motherhood, and creative expression as well as larger questions around the climate crisis and environmental harm, activism and the challenges of working in a neoliberal university environment. Though we came together in the context of a social science writing project we both start from different positions academically—Mila from science, Wendy from the humanities—something which is reflected clearly in the moments of misunderstanding in the dialogue below. Mila is also trained and works in theatre. Wendy has been active in transnational feminism with a focus on body politics as well as feminist political ecology since the late 1980s. There are other traits which mark our perceptions: we are both white cis women. Mila is Dutch and Oxbridge trained. Wendy is from Australia and has lived and worked in Italy and The Netherlands as both an advocate and academic. And, while we both consider our sexuality fluid, we have had the privilege to be able to choose and physically bear children and to raise them with our male partners who are the biological fathers.

Our dialogue is based on three years of discussions as we met in person, individually and in a feminist discussion group, and online through COVID times. If one can speak about a methodology underlying this dialogue, it was about giving ourselves time to listen to each other's approach and be patient with the other's different opinions due to disciplinary assumptions, age and expectations of what an intellectual feminist project could be. We also noted our frustrations at the other's normative assumptions—and this became a key topic in this chapter. Because of the personal impact of our different opinions encountering each other, we also recognised the importance of acknowledging emotions in intellectual debate.

We therefore decided to structure our dialogue around five 'primary emotions': surprise, anger, fear, sadness and joy. As Turner explains, primary emotions are states of "affective arousal" that are thought to be inherent in human neuroanatomy (Turner, 2007). Because of cross-cultural differences in understandings of feelings and emotional expression, the idea of universal classifications of emotions could be questioned; nonetheless, we use these five headings as a tool to show how a range of emotions play out in academic discourse. Under each heading we list the primary emotion and then as subheadings the secondary emotions, followed by our reflections related to that specific emotion and our interpretation of how it is expressed in population debates.

Surprise

Astonishment, Amazement, Shock, Intrigue

> This critique of overpopulation (...) has seemed like a settled issue in feminist circles. Making Kin Not Population: Reconceiving Generations resurrects overpopulation as a question for feminism. Its authors, Adele Clarke and Donna Haraway, are two legendary, influential, and beloved figures in feminist, cultural, and science studies. Indeed, I remember the informal networks abuzz when they first presented this material at a conference in 2013. It was as if they had declared that they had stumbled on a herd of unicorns! (Subramaniam, 2018)

Mila: When I began my PhD in 2018, the first text I read was the collection of essays edited by Adele Clarke and Donna Haraway that Banu Subramaniam refers to in the quote above. One of the things which struck me straight away is the way Haraway and Clarke described the intense climate of debate on population within feminism. I was aware that the population topic is sensitive to many engaged in feminist thought (Bhatia et al., 2020; Murphy, 2017; Sasser, 2018). Yet I was surprised to learn that it was near impossible to discuss this issue with feminist scholars and activists without quickly encountering open displays of feelings such as indignation, disgust, passion and anger.

As Banu Subramanian (2018) states, before Haraway and Clarke started working on population, the population issue seemed settled within feminism: concern about population numbers was unacceptable. This was the case so in part because "direct environmental impacts driven by human numbers are nearly impossible to tease out because they are not, and never have been, simply biological—they are the result of biological, and political, and economic, and technological, and cultural processes and practices" (Sasser, 2018, p. 150). As such, any attempt to reopen the discussion was very unexpected and cause for upset. At the time, there were feminist scholars writing on population to newly criticise those in other fields of academia and policy making, who were expressing concern about population growth. On encountering resistance to their views, there appeared to be a sense of surprise or disbelief among these scholars. Take for example this quote by Ojeda and colleagues: "What is perhaps most surprising about neo-Malthusian environmental thinking is that, despite trenchant critiques questioning its basic presuppositions, it remains as strong as ever" (Ojeda et al., 2019, p. 4). The scholars expressing this surprise are clearly so convinced by the critiques of neo-Malthusianism that they cannot imagine a well-meaning person could possibly disagree with such critiques. In short, in these debates we see a widespread disciplinary agreement among feminists, to the extent that a leading feminist such as Haraway involving herself in the population discussion could cause much dismay and shock.

One of the most surprising discoveries I made, entering the world of feminist debate, was the ease with which what I took for granted as a knowable physical reality was denied by some feminists writing about population. As someone who has been taught to appreciate the emancipative power of statistics in, for example, public health and environmental conservation, I was shocked to learn that there were whole disciplines with a distrust towards numbers and what they represented. In her book 'Figuring the Population Bomb', Carole R. McCann (2017) states she "understand[s] demographic facts to be the products of population theory, a conceptualization of biosocial reality, not a reflection of it (p. 19)" and that "quantification involves an exercise of power that denies it is any such thing (ibid.)." While I had read, within demography, attempts to complicate the understanding of the practices around, for example, census-making, I had never heard the census itself described as depending on "a particular imaginary landscape of 'human bodies' in 'virtual time' and 'virtual space'" (Curtis, 2002, p. 24). I still do not know what to make of such assertions. I would prefer to live in a world with enumeration practices and censuses rather than one without. For an enlightening example, the work of historian William Coleman (1982) in *Death is a Social Disease*, as cited by Haraway (2018), shows that early population thinking and counting led to a better understanding of the apparatuses of inequality and helped galvanise action on public health in urban eighteenth century France. Similarly, I think about the effects early Swedish census-taking had on death rates in that country. In 1749, influenced by the Enlightenment, Sweden became the first country in the world to establish the regular collection of vital statistics (deaths and births) on a national level. They could thereby obtain reliable data on mortality and causes of death, and this data was used to take key steps in improving the health of its populace (Sundin & Willner, 2007). This way Sweden could anticipate and avoid the human devastation of industrial urban growth as seen in places like the UK (Szreter, 2003). In more recent times, time-use studies by feminists have enumerated the disparities between commitments to care between men and women. So, I wonder why among feminists concerned with population there is so much distrust towards numbers, statistics and calculations. Why are time-use studies, or climate physics, seen as important and reliable, but demographic studies scrutinised in order to point to forms of power and domination which they enable?

When I studied demography in my undergraduate, much of the teaching focused on complicating demography's own enumeration practices. The actual number given to a country's population was seen as a useful best attempt to get to the truth, even if not a reliable truth. I vividly remember a lecture about the troubles of collecting reliable census data in a West-African country. Cultural norms about who is a "son" or a "daughter" made a survey question such as: 'how many female and male children do you have?' inaccurate as people would include as their family's sons or daughters any young people who were important to them rather than their birth son or daughter. Additionally, asking questions about stillbirths or infant deaths (the answers being of key importance to a demographer) was not possible due to the stigma that came with the death of babies in the family. By being aware of these issues, demographers could adjust at least partially to such anomalies, to produce statistics which can help us understand rough population trends in those countries. However, coming into feminist debates on population I have missed an open discussion by scholars concerned with the biopolitics of demographic statistics on the criteria of which numbers to trust. Which numbers should we trust and which not? And why, I wondered, is this all so controversial?

I've since learned more about the biopolitical controversies around the topic of population, particularly concerns around racism and colonialism. Philosophical and practical questions around human numbers are fraught with ideological differences. As Diane Coole argues, affecting fertility rates is 'profoundly controversial'. At stake are "liberal values of freedom, autonomy and human rights, entangled here with contested definitions of sexuality, gender roles and identities, family norms and embodiment, as well as with ideological disputes over the role of the state and its powers" (Coole, 2018, p. 4). With a topic that touches so many foundational political and personal questions, no wonder there is so much disagreement, especially across disciplines. The bigger question then becomes: how to manage the conflict and resulting emotions in population debates?

Wendy: First, I admit I was surprised at your comment that there is a knowable physical reality that is captured in demographic studies and cannot be challenged as there is a truth to how we measure populations. My Foucauldian training in bio-politics pointed to the colonial roots of population statistics particularly as used in wide sweeping global population studies, which used numbers to obfuscate historical economic, cultural and political oppression. The concerns around Malthusianism in demographic debates continues. Betsy Hartmann's (1995) *Reproductive Rights and Wrongs: The Global Politics of Population Control* is still a classic in feminist studies on reproductive justice. Her work scrutinises the use of population numbers in discussions of environmental and intergenerational justice. She argues that reproductive justice must be based not on statistics but on understanding social processes and institutions which create communities and provide the social, economic and ecological conditions that support human security and sustainability upon which, ultimately, all production, exchange and accumulation rest. The fight for reproductive justice is not about how many children are born into families and how many people are dying, but also about social, economic, civic and environmental goals. These concerns scrutinise population studies in struggles against patriarchy, racism, classism and extractivism. Feminists like Hartmann, Sasser and Ojeda are deeply concerned about renewed use of statistics in the climate debate which provide the ongoing justification for the control of racialised bodies in population policy (Hartmann, 1998; Ojeda et al., 2019; Sasser, 2014). I am sympathetic to their exposure of these struggles and their call for diverse strategies to build decolonized, socially just futures. However, I do see your concern around the othering and silencing that goes on among feminists and environmentalists and the quarrel around numbers in the population debates.

So, indeed, I was surprised by Haraway and her call for "making kin not babies". What I see as positive in the debate is that Haraway asks feminists to consider new forms of knowledge which value kin—other-than-human life—as part of the feminist project to unpack corporate power, technoscience and biopolitics. The invitation is to forge a multi-species eco-justice that breaks through gendered and racialised nature of biology, culture and technology.

What intrigued me most was Haraway's concept of speculative fiction and how to engage our imaginations in thinking about how human and other-than-human lives need to be considered as kin. I found this idea to be unsettling. Like her Cyborg Manifesto, which we discuss below (Haraway, 1990), Haraway provokes our feminist imaginations and our feminist politics. Social science fiction, the art of telling stories, and going beyond the apparent scientific givens of reproductive bodies are appealing to me as someone trained in humanities. In reading Haraway and eco feminists such as Val Plumwood (1993), environmental humanities scholars such as Deborah Bird-Rose (2013) and Indigenous writers such as Aileen Moreton-Robinson (2000) we are challenged to see all 'earthlings' as kin and the need for better care for all, including being able to mourn the losses and destruction. Acknowledging the need to care for more than human others deeply resonates with my desire for 'understanding otherwise' and decentring humans from our understanding of eco-justice. It seems liberating that we could consider multiple kin as part of our battle for reproductive justice. This helped me go beyond the human numbers game and move towards thinking about our responsibility as humans to other beings on the planet.

Haraway is asking feminists to reimagine kinship, family, and reproduction and to talk about the politics of reproductive justice and our complex relations to others. Reproductive justice is about sustaining the conditions necessary for collective thriving—including environmental justice, food justice, climate justice, antiracist social justice—and nonviolent ways of relating to human and more than human others. Her retelling of reproductive justice in the future with social science fiction stories or 'narrative speculative fabulation' about future technologies that merge human and more-than-human forms are indeed fabulous in her tentacular thinking, (see for example the story of the Chthulucene as an alternative to the concept of the Anthropocene, Haraway, 2016, p. 55).

Anger

Exasperation, Frustration, Resentment, Disgust, Indignation, Annoyance

> I have been screamed at after lectures by my feminist colleagues of many years, told that I can no longer call myself a feminist (...) for arguing in public that the weight of human numbers on a global scale, however broken down by analysis of structured inequalities, opposition to ongoing racist population control programs, and many other important things, is an outrage. (Haraway, 2018, p. 87)

Mila: As someone who loves fiction, I appreciate what you say about Haraway's Speculative Fabulation (SF). However, in this case my main intellectual interest remains very pragmatic; my focus is on the clashes of opinion between academics and the emotions which population discourses bring up. Ultimately, my PhD research led me to believe that interpersonal and intergroup dynamics have a profound impact on the knowledge about population which is created and put forward to, for example, policy makers. In 2018 and 2019, I attended four different reading groups on the *Making Kin Not Population* at two different universities in the UK. I ended up coming away each time with a realisation that the topic of the discussion was too inflammatory to lead to in-depth discussions. Many of the comments made related to the emotional responses of the readers, not their thoughts, arguments or intellectual engagement regarding population. Someone said: "This book left me infuriated for weeks." Someone else jokingly proposed imagining a street fight with team Murphy (Haraway's co-author Michelle Murphy, who argues against the use of the word population altogether) and team Haraway. "Yeah, we'd have t-shirts for each!" They then asked the group: "Who are you with?" Various people emphatically said, "I am Team Murphy." Not one person dared to say, "I am team Haraway", though a few people stayed quiet. The joking didn't last long. Reflecting on it later I realised I felt that the general mood of the discussion was that of frustration and indignation towards Haraway. While there was some interest in the 'Making Kin' part of Haraway's slogan, the content around 'Not Babies' was, I would say, entirely ignored.

Before this intervention by Haraway, academic feminism as an activist field of scholarship, had clearly settled on a certain set of norms around population. From what I could see, students are not only trained in the debates on population and its history, but also on what is acceptable to say about reproduction and population. There is a sense that in feminist circles it is not socially safe to express any doubt or concern about population growth. Instead, related concerns are quickly moved to questions about sustainable consumption or to problems such as eugenics and colonialism that are inherent in much of the historic elements of population control. Deviation from this norm seems to inspire anger.

I do believe that this comes from the best of intentions and a genuine belief in the harms of populationism/neo-Malthusianism. The following anecdote in a podcast is very telling. In 'Imagine Otherwise' with Cathy Hannabach (2019) Jade Sasser—a scholar who focuses on gender, climate justice, and reproductive politics—tells us about a frustrating teaching experience she regularly has:

> But with that said, what also happens every quarter in the classroom that is intensely frustrating to me because I don't know what to do about this, is that students will hear me spend an entire hour and a half or even several weeks offering a very critical, very nuanced, very challenging perspective on population control. Then after all that, they'll still go back to, 'Well, but we need to slow or control or end population growth because of climate change. Population is still something that we need to really tackle because of climate change.' And I'm like, 'Have you not listened to everything I've been saying?' What I'm doing in my work is, I'm really trying to disrupt and dislodge paradigms, knowledge paradigms, and it's hard for young people to let those paradigms go because they've been raised with them. But I continue to persevere. I won't give up. I will continue to challenge my students' thinking and really try to disrupt and dislodge the idea that population control is a natural and necessary component for environmental conservation.

I sympathise very much with Sasser's sense of exasperation. I learned a lot from her thorough work in *On Infertile Ground: Population Control and Women's Rights in the Era of Climate Change* (2018). However, I can't help but think that her "have you not listened to everything I just said?" could also be spoken to her—exclaimed even—by some of the equally nuanced and careful scholars who do warn about potential hardship caused by growing human numbers. As far as I can see there are very valid points made by people who call each other opponents, enemies even (see Wendy's point below). Are Sasser's students not continuing to press their own population concerns after hearing the critiques because they also come across other positions, which convincingly present the inevitable physical reality that the numbers *are*, at times, problematic? Could that not be the same reason that those who are (informally) identified by some feminist scholars as 'being in the enemy camp'—some demographers, certain environmentalists—are continuing to express concern at growing populations, albeit in smaller numbers than in the twentieth century?

Even just entering this debate I could find myself getting frustrated. Not with one party or the other, but with the lack of actual content-based dialogue across disagreement. Why is it so hard to see different types of population knowledge as merely partial truths, as needing synthesis? To my annoyance, I observe a lot of ignoring of the others' arguments, on both sides, and this seems like an utter shame. It seems to me scholars working on reproductive rights, population and environmentalism often find themselves in bubbles and are not listening to the nuances of those in other areas. Or that the anger and indignation is so strong that even when someone like Haraway is making a considered plea for dialogue, she is met with fury by some. I want to say: 'Just think with her! Being in dialogue does not mean agreeing!'.

Wendy: I am sympathetic with your frustration and strong feelings which come out when you speak of 'utter shame'. There is anger, even despair when we engage in debates where people do not listen. So much anger erupts around sexual health and reproductive rights which underline population debates about 'family planning' and contraception. I have witnessed over the years tense and loud arguments in UN meetings when representatives of the Catholic Church and other conservative groups would move into rooms and start disrupting discussions. I recall being in a room of a high-level UN official as he was listening to the Vatican Radio decrying the latest World Report on Population (which he edited) and his disgust at what was being said, knowing he would have to face them down in future meetings that would be deciding a country's health budget based on concerns around whether money would be used for sexual health needs. During UN meetings in the 1990s, I would band together with other feminist advocates in the different regional SRHR movements and NGO networks order to plan strategically our interventions and speeches knowing there would be a right-wing attack to confront from conservative and religious NGOs and governments- and that we would have to battle for every word that touched on sexuality or women's right to choose.

In academe I have seen less room for anger to be expressed directly. There is, though, often a sense of indignation that scholars can feel about their work not being considered or heard by those other academics that do not share their views. I have personally felt considerable indignation at how feminist political ecology perspectives are ignored by political theorists and economic scholars. To take as an example, I reviewed a recent book by Sir Partha Dasgupta on population ethics—a branch of moral philosophy (guided by economic and climate science) that looks at how the numbers of people impact the quality of life of others in the future (Harcourt, 2020). His book totally ignored feminist or gender debates, so I was literally gritting my teeth when Dasgupta states he is "just trying to get the numbers right … nothing more" (Dasgupta, 2019, p. xxxiii). I felt angry at his dismissive 'back of the envelope' empiricism as he asks 'birth and death' questions which touch major concerns around gender, reproductive decision making and natural resource use as well as a host of other socio-ecological concerns without acknowledging the context. As an advocate I have approached the questions of population, consumption and environment from a critical gender, development, and human rights perspective, engaging in transnational advocacy and policy work with organisations such as the UNFPA and the World Health Organization. I therefore resented Dasgupta's lofty tone as he uses esoteric models to tell us "how to study the population–consumption–environment nexus, in order to tell us how far we are today from where we probably should be" (Dasgupta, 2019, p. 218). And then, as a feminist political ecologist, I felt indignant that he refers to deep emotional needs that 'we' all have to create children and then the unbearableness of life for the half a billion people who are malnourished and prone to disease, living in conditions where 'you' wouldn't want to create children.

His work is at completely at odds with my feminist political ecology approach which scrutinizes the use of population in discussions of environmental and intergenerational justice particularly around social reproduction as "social-environmental process required to maintain everyday life and to sustain human cultures and communities on a daily basis and intergenerationally" (Di Chiro, 2008, p. 281). Dasgupta's neoclassical modelling erases the entangled relationships between population growth and environmental problems. Why, I thought while writing the review, is population ethics determined by views such as Dasgupta's, seen through the prism of economic modelling about the 'right numbers of people', rather than seeing how societies and therefore economies are embedded within nature? Do numbers matter? Have we actually tried to live sustainably – not just live differently (as in our everyday habits), but also organise our societies differently and do politics differently, so as to collectively address environmental degradation and inequality without having to impose reproductive decisions?

As we discussed together, quite heatedly, numbers can manipulate and obfuscate. At one point you asked me to look at the 2010 Ted Talk of the Swedish academic Hans Rosling, where he uses the story of his mother's washing machine to discuss the thorny question of how to distribute the world's resources so people can benefit from using washing machines without destroying the planet. The story he told, we realised, helped us understand we were not so different in our concerns. For me the issue was not about 'how many people' but about distribution and justice. The story Rosling was telling was not about numbers, per se, but about everyday lives, technology, gender, work and global inequities. If the carrying capacity is 3 or 11 or 20 billion people is not the point- far more powerful and important are the multiple and complex interrelationships that raw numbers alone can obfuscate.

Fear

Alarm, Apprehension, Hysteria, Horror, Panic, Nervousness, Uneasiness

> I think that is part of the problem 'we' face. The subject is forbidden, no matter how carefully it is framed; it has been ceded to the right and to population professionals. To insist that seriously facing the burden of human numbers is not racist; but shutting up out of terror of the issue might well be. Fear of getting things badly wrong certainly doesn't serve reproductive justice, even in human-exceptionalist terms, much less in terms of multi-species reproductive justice. (Haraway, 2017).

Mila: Rosling's storytelling is wonderful, and indeed we learned through it that you and I both cared for many of the same elements of possible human flourishing. Yet Rosling gave that talk when population projections were far more optimistic (in my view) than they are now. I sometimes find myself fearful, along with some demographers, that if population in specific places continues to grow with the speed they are now – in the context of the capitalism-driven unequal world we live in – it will lead to further immense poverty and societal disruption.

To be specific, while I do not care about the absolute nature of aggregate population numbers, I worry very much about the *speed* with which some populations are growing. For me this is not about being racist but about recognising that this rapid population growth can lead to major problems in timely infrastructure creation, increasing global inequality and vulnerability for many (Coole, 2018; Rougoor & van Marrewijk, 2015; UN Department of Social Affairs, 2011). And that reducing the rate of population growth could make it easier to address existing problems. For these reasons I would say that a good analysis of the potential impact on a particular society of (1) rapid population increases and (2) the presence or absence of a population programme, requires both numbers *and* qualitative research that address multiple and complex interrelationships. The same holds for policies in this area—they need both types of work, both lenses. In short, I do not believe numbers *necessarily* obfuscate other essential considerations, even if they sometimes do.

Wide-spread provision of voluntary family planning services and educational opportunities for girls and young women are the policies which most twenty-first century population control advocates call for. I think you and I agree that these are important things in themselves. Where we may differ is that I also think that it is okay to try to get more funding for them by leveraging population concerns. And that I think it is important that people and governments are well-informed of the concerns of demographers and others who believe there might be trouble ahead if birth rates in certain places are not reduced quickly—as well as of the concerns of those who centre SRHR and oppose (aspects) of population policy. Due to the problematic history of population and the forceful arguments in the feminist discourse on population, concern about population has become somewhat of a taboo in many areas (Campbell, 2007; Coole, 2013; Singer & Kissinger, 2017). This has meant that there are fewer academics and NGOs working on this question than there were previously (Mora, 2014) and such knowledge may not reach the places where it could be relevant.

On a different type of fear: I am still convinced that population debates do not currently address the arguments in themselves but are mired in name-calling—a typical response derived from fear (Shapiro, 2010). Changing your opinion on important matters can be frightening. So, it is sometimes easier to simply exclude the possibility that one is wrong and instead push away anyone who argues to the contrary. Political ecology teaches us to question dichotomies of 'right' and 'wrong' and promotes the value of bringing together different methods, data interpretations and opinions. I think viewing the population debate with such a lens would go a long way in communicating more fruitfully across difference.

In the case of Haraway, dismissing her and her new ideas was difficult, because of her unique position as a well-loved feminist, a giant in the field. Here Haraway is accused of being genocidal and anti human but also there are attempts to—as it were—split her public persona into different parts: A Haraway to love, one who gave us idea of the "the God trick" (1988) and "cyborg imagery" as "a way out of the maze of dualisms in which we have explained our bodies and our tools to ourselves" (Haraway, 1990, p. 316). And then a Haraway to ignore, a Haraway who must somehow be mistaken, must have lost her intellectual acumen or her revolutionary spirit.

Yet I think Haraway's work on population is evidence of the same intellectual fearlessness as she displayed by writing the Cyborg Manifesto. I suspect she is very aware of her position in this debate, of what is and what is not expected in terms of population opinions and what effect her particular prominent position has on the arguments she makes. She was always a provocateur, no? The Cyborg Manifesto is now widely loved by feminists, but at the time, wasn't it a very prickly set of arguments she made?

Wendy: While I am not in agreement that the focus on population numbers helps us deal with inequality and that population debates are devoid of colonial and racist views and othering, I do agree that the Cyborg Manifesto was a major intellectual intervention in feminist science and technology studies and was in a sense fearless in its critique of ecofeminism, and a feminist championing of science and technology studies. She looked squarely at the fear that technologies would invade our personal and intimate lives. In breaking through those fears, Haraway's Manifesto helped us to look at how technology is infused with the political, cultural and material embodied experiences.

Mila: Wait, I may not have been clear enough up to this point. You say you disagree with me and think that I don't see how population debates are colonial and filled with racist views and othering. However, I in fact do agree that the history of population discourse and practice is linked to colonialism and class and race-based prejudices. What I do not agree with is that population debates – past, present and future – are *necessarily* racist, classist, colonial and so on. For me the distinction between 'often is/has been racist' and 'is not necessarily/can be not racist' is key. I would like to promote dialogue across disciplines and viewpoints so that, for example, anti-racism *and* population concerns can come together to promote better policy. This requires a recognition of the historic injustices which have been perpetuated in the name of population. But it also requires that population policy responds to real-life issues for many people, including people of colour in the Global South. I would argue that population discourse and policy frameworks can be developed within a wider progressive politics in ways that can become a force for good.

Wendy: Your interruption is valid, particularly as you point to the need for us to listen to each other. However, we are still not in full agreement. For me progressive politics should be about distribution of resources, changes in rich people's lifestyles and openness to all women's reproductive choices, full stop, not a set of 'population' policies that aim to reduce numbers because too many (poor) people cause too much environmental damage. Diversity and context matter, and who can access what and who is deciding who (else) is too many. I remain worried about how this is all playing out. I fear that the likelihood of regular global pandemics and heat waves, fires and floods due to climate change will produce an inequitable set of policies if we do not point to racial, colonial and patriarchal discourses underlying current policies and seek very different behaviour (not just good policy). In fact, climate impacts are already affecting the poorest countries, marginalised people, and racialised people.

To return to cyborgs, well we are not so fearful of cyborg life as it is now normalised. This is one reason why I see Haraway's idea of 'making kin not babies' as full of possibilities. Nevertheless, it is with a sense of uneasiness that I take up this call to make kin not babies (or population). As I found in Dasgupta's text, academics and government workers who are engaged in population policy typically do so from different angles than Haraway's creative way of helping us envision futures. Most demographers do not see themselves as storytelling but as empirically telling the truth when discussing changing population trends and patterns and the policies required to reduce population numbers to conserve the environment. Talking about social science fiction, writing manifestos, describing personal stories are not usually acceptable academic truths to the majority of demographers.

And at times, I admit, I do give into fear and lose hope. I am afraid of the continued violence, not only the current escalation of wars, the femicides, deaths of environmental defenders, journalists, but also the dark worlds of Internet gaming etc. Such 'naturaltechnical' worlds are a far cry from Haraway's speculative fiction where human genes mix with butterflies. How do we rethink kinning in a world dominated by such oppression, violence and uncertain futures? What we can learn from Haraway is that the question is not a yes or no to technology "invading" life, or a yes or no to having (more or less) kids, but what are the surrounding ethics that we must cultivate to inform such decisions. It is one thing to decide not to have children because I want to make kin with my neighbours, my dog, the sea. It is another to be coerced into not having children for the good of biodiversity. We need to give attention to the ethics and politics that informs the different debates around population and socionatures/naturecultures.

Sadness

Disappointment, Shame, Grief, Despair, Gloom, Isolation, Rejection, Dejection, Guilt, Regret

> This is a brazenly personal paper and a plea for other-than-biogenetic kindred. I begin with a painful mass in my gut, pressing up against my diaphragm until it ruptures. The pain is much like the bodily feeling of grief when my mother died, when my first husband died, when my father died, when the dog of my heart died - the feeling of grief, exploding from the inside out, evisceration, terror. (...) But the pain I feel in my belly has to do with something else (...) the surplus killings of ongoingness, the wanton surplus extinction of kinds, of whole patterns of living and dying on earth, of genocides across human and other than human groups. (Haraway, 2018, p. 69)

Wendy: Like fear, sadness is so much part of our lives right now. You try to shake it off but it is difficult. Perhaps it is closer to despair as Haraway so viscerally and powerfully expresses in the quote above. I look back over the years and wonder at why we are in the crisis (crises) we predicted when I was a student feminist and environmental activist over 30 years ago. Why is it so hard to get out of systemic violence even if it is being named and discussed all the time? Is it just because I exist in my small bubble - even if it is a transnational bubble and one that stretches over decades full of exciting conversations and what looked like contributions to transformative change? What is my responsibility for the failures? Individually I seem to have benefited well enough from this deeply unfair world.

However, I continue to engage and be inspired—from the courage of others and their stories in end times. I feel it is important to learn as feminist academics to value ways to communicate differently, using art, film, theatre, murals, creative spaces to allow our imagination to be positive, and see that as knowledge alongside the positivists' 'truths'. It is not for me about reducing numbers, but about taking up responsibility which is not just about providing contraceptive choices. It seems more complex than this as I read, watch TV or doomscroll on my phone about the increasing level of violence, war, extractivism and toxic pollution, and recognise my awkward place in the racialised violence of modernity that has benefited me personally at the expense of others. Even if I celebrate some of the changes for some women's lives and their choices, I still feel despairing at what is not happening, from the lack of contraceptive choice and the increase in sexual violence to the oppression against peoples who do not conform to heterosexual norms, to the erasure of cultures and the overwhelming loss of biodiversity and beauty in nature. I remember the first time I heard about tipping points, now 15 years ago, from a biologist and feminist friend. I cried then. But I couldn't believe I would live to see so many tipping points smash bang in our face.

My sadness extends to when I hear so many young people questioning if they should have children as they face economic uncertainty and consider the devastating impacts of climate change in these end times. We need to be aware of a creeping individualization of responsibility which is capturing environmentalism. Deciding to have children when I did was so much more about my choices. I thought I was fighting for the individual choice to have children (the biological, technological and economic choice). Now that 'choice' has become much more entangled in social and environmental responsibilities which diminish the possibility of the individual to speak unaware of collective responsibilities and fearful futures.

Joy

Relief, Hope, Eagerness, Enthusiasm

> What if making a baby became truly an act of joy and material, daily responsibility for an enlarged community? How to celebrate children in non-natalist movements? (...) How to celebrate human maturity for women and men in building selves and communities without making babies? (Haraway, 2018, p. 97)

Mila: I also see many of my peers choosing not to have children. Some because of the life(style) they envision for themselves, but many also because of fear of a climate catastrophe. And some worry about the culpability of bringing into the world another European human who will consume and pollute the environment 80 or 90 years ahead in time. I asked myself about these issues when I came to the decision to try to have children or not. At the time, the connection I felt to my partner and my desire to create new life, a family, with him, – for me the ultimate commitment – was a far more convincing future than the one in which I saw my offspring as a planetary liability. I also thought back to how my parents in the late eighties were told by their friends that they were mad to try to bring a baby into the world; after all, the nuclear bomb could drop any moment and bring global devastation. Their child might only ever know great suffering. My parents, living in a squatted farmhouse, without secure jobs, still young, decided against acting on that fear. Lucky for me! I was born, followed by my sister and brother a few years later. And the question of my own reproduction and the risks this would bring for my then hypothetical children came down to a simple comparison: I am so glad to be alive, so grateful I get to be here to experience human existence with all its confusions and pleasures, that I expect that my children might well come to feel the same. They will, however, have to face ecological and climate breakdown and all the unprecedented and incalculable societal changes that will come with that. Perhaps then this is one of my primary tasks as a parent: help cultivate in my children the ability to experience joy, no matter what the circumstances, also in the face of suffering. For now, I am simply so very glad they are here with me on this planet, and as far as I can tell from their endless vigour and frequent laughter, so are they.

Wendy: Such a beautiful birth story Mila. I too can speak of joy and hope, individually as a mother, and collectively as part of communities who help me to find ways to relate and sustain ourselves, our kin and our environment. For me this joy is always mediated as I continually negotiate social practices of mutual support that enable strategies of living well together. The different feminist communities, whether they have been activist, academic or friendship based (and sometimes all three), have enabled me to flourish and enjoy life in deeply important ways. In the last few years, I have tentatively begun to acknowledge my joyful relationship with different environments that support and sustain me. Whether they are the oceans in Australia or the lakes in Italy or the woods in The Netherlands or even the plants and flowers that grow on my terrace, some that have been gifted to me by students over the years., I certainly feel joy in these living beings. This is kinning, as their presence offers a tiny but sustaining way to continue facing the overwhelming concerns of our times.

Conclusion

> In a political and cultural moment where debate is enacted through name-calling, slander, falsehood, and labelling ones' opponents as treasonous enemies, I am deeply moved by this collection [the Making Kin Not Population book]. Some of us will never agree, but the book reminds us of the critical need to engage rather than disengage, and to argue respectfully rather than blame or ignore those we disagree with. (Subramaniam, 2018)

By boldly using emotions as a way into the tricky topic of population guided by Haraway's invitation to 'make kin not babies' we have tried to listen constructively to each other and to those with whom we did not agree rather than draw up camps of us and them. We have had time to build up enough trust to pause and listen to each other when we started to note emotions rise. We had to listen hard to what the other had to say about numbers and fearful futures. In those moments we slowed down and chose every word carefully so we would not be misconstrued. We also recognised as we spoke that there were taboo topics, stories that could not be told in an academic text, as they were too sensitive or would evoke dismissal. In other projects, where we went beyond the written word, to theatre performance (Mila) or art (Wendy), we recognised we could allow more to be expressed, understood and heard.

We have aimed to open up the debate on population in FPE not only by introducing Haraway's idea of 'making kin' as a feminist strategy for survival, but also by paying attention to the emotions in which the debate is couched. We have noted the different disagreements within feminist circles as well as between feminist and environmentalists as well as Haraway's acknowledgement of the negative responses she received from colleagues. Haraway's descriptions of emotion in these discussions may well have been part of her own rhetorical device to position her in the debate, but they do foreground how much emotions shape academic debate. Paying attention to the role of emotions in academic work adds complexity to the debate but also can propose ways to break down taboos and open up constructive discussion.

In our conversations, it occurred to us that the large disagreement on the best way to consider population growth might obscure the fact that many scholars engaged in population debates share a similar goal: to influence policy to improve the well-being of people now and in the future. Taking this as an explicit starting point when engaging with those we otherwise disagree with could go a long way in allowing constructive dialogue to develop.

Funding: This chapter was funded by the Wellbeing Ecology Gender and cOmmunities Innovation Training Network (WEGO-ITN) funded by the European Union's Horizon 2020 research and innovation programme under the Marie Sklodowska-Curie grant agreement No. 764908-WEGO 2018-2021.

References

Barney, J. B. (1990). The debate between traditional management theory and organizational economics: Substantive differences or intergroup conflict? *Academy of Management Review, 15*(3), 382–393. https://doi.org/10.5465/amr.1990.4308815

Bashford, A. (2014). *Global population: History, geopolitics, and life on earth*. Columbia University Press.

Bhatia, R., Sasser, J., Ojeda, D., Hendrixson, A., Nadimpally, S., & Foley, E. E. (2020). A feminist exploration of 'populationism': Engaging contemporary forms of population control. *Gender, Place & Culture, 27*(3), 333–350. https://doi.org/10.1080/0966369X.2018.1553859

Bird-Rose, D. (2013). Val Plumwood's philosophical animism: Attentive interactions in the sentient world. *Environmental Humanities, 3*(1), 93–109.
Campbell, M. (2007). Why the silence on population? *Population and Environment, 28*(4/5), 237–246.
Charmaz, K. (2012). Writing feminist research. In S. Hesse-Biber (Ed.), *Handbook of feminist research: Theory and praxis* (pp. 475–494). SAGE Publications, Inc. https://doi.org/10.4135/9781483384740.n22
Clarke, A. E., & Haraway, D. (2018). *Making Kin not population: Reconceiving generations*. Prickly Paradigm Press. https://www.press.uchicago.edu/ucp/books/book/distributed/M/bo28583407.html
Coleman, W. (1982). *Death is a social disease: Public health and political economy in early industrial France*. University of Wisconsin Press.
Collins, R. (2000). *The sociology of philosophies: A global theory of intellectual change* (New ed.). Harvard University Press.
Collins, R. (2002). On the acrimoniousness of intellectual disputes. *Common Knowledge, 8*(1), 47–70. https://doi.org/10.1215/0961754X-8-1-47
Coole, D. (2013). Too many bodies? The return and disavowal of the population question. *Environmental Politics, 22*(2), 195–215. https://doi.org/10.1080/09644016.2012.730268
Coole, D. (2018). *Should we control world population?* Polity Press.
Curtis, B. (2002). *The politics of population: State formation, statistics, and the census of Canada, 1840–1875*. University of Toronto Press.
Dasgupta, P. (2019). Time and the generations: Population ethics for a diminishing planet. *Columbia University Press*. https://doi.org/10.7312/dasg16012
Di Chiro, G. (2008). Living environmentalisms: Coalition politics, social reproduction, and environmental justice. *Environmental Politics, 17*(2), 276–298.
Elmhirst, R. (2011). Introducing new feminist political ecologies. *Geoforum, 42*(2), 129–132. https://doi.org/10.1016/j.geoforum.2011.01.006
Halfon, S. (2006). The disunity of consensus: International population policy coordination as socio-technical practice. *Social Studies of Science, 36*(5), 783–807.
Hamilton, J. M. (2017). The trouble with babies [Book Review]. *Cultural Studies Review, 23*(1), 185.
Hannabach, C. (Host) (2019, February 27). *Imagine otherwise: Jade S. Sasser on reproductive justice and climate change* (No. 83) [Audio podcast episode]. In Ideas on Fire. Retrieved November 26, 2021, from https://ideasonfire.net/83-jade-sasser/
Haraway, D. (1988). Situated knowledges: The science question in feminism and the privilege of partial perspective. *Feminist Studies, 14*(3), 575–599. https://doi.org/10.2307/3178066

Haraway, D. (1990). A manifesto for cyborgs: Science, technology, and socialist feminism in the 1980s. *Feminism/Postmodernism, 190*, 149–181.

Haraway, D. (2015). Anthropocene, capitalocene, plantationocene, chthulucene: Making Kin. *Environmental Humanities, 6*(1), 159–165. https://doi.org/10.1215/22011919-3615934

Haraway, D. (2016). *Staying with the trouble: Making kin in the Chthulucene*. Duke University Press.

Haraway, D. (2017). Response to life with Ms Cayenne Pepper. *London Review of Books, 39*(13). Retrieved 28 July 2022 from https://www.lrb.co.uk/the-paper/v39/n13/letters

Haraway, D. (2018). Making Kin in the Chthulucene: Reproducing multispecies justice. In D. Haraway & A. Clarke (Eds.), *Making Kin not population: Reconceiving generations* (pp. 67–100). Prickly Paradigm Press. https://www.press.uchicago.edu/ucp/books/book/distributed/M/bo28583407.html

Harcourt, W. (2009) *Body Politics in Development: Critical debates on gender and development*. Zed Books.

Harcourt, W. (2020). Unravelling the 'P' word in environment and development. *Development and Change, 51*(6), 1628–1639. https://doi.org/10.1111/dech.12619

Harcourt, W., van den Berg, K., Dupuis, C., & Gaybor, J. (Eds.). (2022). *Feminist methodologies: Experiments, collaborations and reflections*. Springer International Publishing. https://doi.org/10.1007/978-3-030-82654-3

Hartmann, B. (1995). *Reproductive rights and wrongs: The global politics of population control*. South End Press.

Hartmann, B. (1998). Population, environment and security: A new trinity. *Environment and Urbanization, 10*(2), 113–128.

Hawkins, R., Ojeda, D., Asher, K., Baptiste, B., Harris, L., Mollett, S., Nightingale, A., Rocheleau, D., Seager, J., & Sultana, F. (2011). A discussion. *Environment and Planning D: Society and Space, 29*(2), 237–253.

Lewis, S. (2017, May 8). *Cthulhu plays no role for me*. Viewpoint Magazine. Retrieved July 28 2022 from https://www.viewpointmag.com/2017/05/08/cthulhu-plays-no-role-for-me/

McCann, C. R. (2017). *Figuring the population bomb: Gender and demography in the mid-twentieth century*. University of Washington Press.

Mehta, L., Huff, A., & Allouche, J. (2019). The new politics and geographies of scarcity. *Geoforum, 101*, 222–230. https://doi.org/10.1016/j.geoforum.2018.10.027

Mollett, S., & Faria, C. (2013). Messing with gender in feminist political ecology. *Geoforum, 45*, 116–125. https://doi.org/10.1016/j.geoforum.2012.10.009

Mora, C. (2014). Revisiting the environmental and socioeconomic effects of population growth: A fundamental but fading issue in modern scientific, public, and political circles. *Ecology and Society, 19*(1). https://doi.org/10.5751/ES-06320-190138

Moreton-Robinson, A. (2000). *Talkin'up to the white woman: Indigenous women and feminism*. University of Minnesota Press.

Morgan, M., & Baert, P. (2015). *Conflict in the academy: A study in the sociology of intellectuals*. Palgrave Macmillan.

Murphy, M. (2017). *The economization of life*. Duke University Press.

Ojeda, D., Sasser, J., & Lunstrum, E. (2019). Malthus's specter and the anthropocene. *Gender, Place & Culture, 27*(3), 316–332. https://doi.org/10.1080/0966369X.2018.1553858

Plumwood, V. (1993). *Feminism and the mastery of nature*. Routledge.

Richardson-Ngwenya, P., & Nightingale, A. J. (2018). Diverse ethics for diverse economies: Considering the ethics of embodiment, difference and inter-corporeality at Kufunda. In C. Bauhardt & W. Harcourts (Eds.), *Feminist political ecology and the economics of care* (pp. 131–161). Routledge.

Rocheleau, D., Thomas-Slayter, B., & Wangari, E. (1996). A feminist political ecology perspective. In D. Rocheleau, B. Thomas-Slayter, & E. Wangari (Eds.), *Feminist political ecology: Global issues and local experiences* (pp. 3–26). Routledge.

Rougoor, W., & van Marrewijk, C. (2015). Demography, growth, and global income inequality. *World Development, 74*, 220–232. https://doi.org/10.1016/j.worlddev.2015.05.013

Sasser, J. (2014). From darkness into light: Race, population, and environmental advocacy. *Antipode, 46*(5), 1240–1257. https://doi.org/10.1111/anti.12029

Sasser, J. (2018). *On infertile ground: Population control and women's rights in the era of climate change*. NYU Press.

Schultz, S. (2021). The neo-Malthusian reflex in climate politics: Technocratic, right wing and feminist references. *Australian Feminist Studies, 36*(110), 485–502. https://doi.org/10.1080/08164649.2021.1995847

Shapiro, D. L. (2010). Relational identity theory: A systematic approach for transforming the emotional dimension of conflict. *American Psychologist, 65*(7), 634–645. https://doi.org/10.1037/a0020004

Singer, P., & Kissinger, F. (2017, August 6). *Rethinking the Population Taboo*. Syndicate Project. Retrieved 28 July 2022 from https://medium.com/project-syndicate/rethinking-the-population-taboo-f68d53ffb085

Subramaniam, B. (2018, November 27). *"Overpopulation" is not the problem*. Public Books. https://www.publicbooks.org/overpopulation-is-not-the-problem/

Sundberg, J. (2004). Feminist political ecology. In D. Richardson (Ed.), *International encyclopedia of geography: People, the earth, environment and technology:*

People, the earth, environment and technology (pp. 1–12). Blackwell & Association of American Geographers.

Sundin, J., & Willner, S. (2007). *Social change and health in Sweden.* Swedish National Institute of Public Health.

Szreter, S. (2003). The population health approach in historical perspective. *American Journal of Public Health, 93*(3), 421–431.

Torracinta, S. (2017, September 19). Make Kin, not borders. *The New Inquiry.* Retrieved 28 July 2022 from https://thenewinquiry.com/make-kin-not-borders/. Accessed 28 July 2022.

Turner, J. (2017, June 1). Life with Ms Cayenne Pepper [Book Review]. *London Review of Books*, 23–27.

Turner, J. H. (2007). *Human emotions: A sociological theory.* Routledge, Taylor & Francis Group.

UN Department of Social Affairs. (2011). *United Nations, World Population Prospects: The 2010 Revision.* Population Division, United Nations.

CHAPTER 11

La Mercadita 2050: Telling Tomorrows of a Market After Oil

Lillian Sol Cueva

Introduction

Imagine waking up tomorrow in a city that does not depend on fossil-fuels. What kind of energy will we have? Which energies will power economic and social systems, and how do they shape our lives? Also, picture how we mobilize through that future. Ask yourself, whose voices were listened to in order to reach that future and what, if any, visions were hegemonic?

When I try to answer these questions, I see two opposite scenarios informed by my interest in energy transitions and my political engagement as a feminist. On the one hand, I see a single future in which more technologies are put in place, but not so much has changed in terms of equality, justice, or careful relationships with humans and more-than-humans. On the other hand, I envision a plurality of futures beyond

L. Sol Cueva (✉)
International Institute of Social Studies, Erasmus University of Rotterdam, Rotterdam, The Netherlands
e-mail: solcueva@iss.nl

W. Harcourt et al. (eds.), *Contours of Feminist Political Ecology*, Gender, Development and Social Change,
https://doi.org/10.1007/978-3-031-20928-4_11

exploitation; where less energy is needed, and everything is powered by a mix of renewable and human-powered kinetic energy. In these futures, my voice and other women's voices are listened to, and they shape the communities in which we live in.

Yet, while trying to answer the above questions, I realize that there are many more imaginaries[1] than my utopian or dystopian ones of how the future will be. It makes sense to ask wider questions about the future and to ask them to the people who have not been listened to by experts—governments, energy industry, technicians—in order to speak in the plural of futures and to open up the possibility of imagining and envisioning other realities. After all, dominant and alternative imaginaries have the potential of limiting pathways by closing down alternatives that individuals or social groups can choose from, or they have the capacity of opening up pathways that can challenge the vision of the powerful (Birch, 2017).

In Mexico, fossil-fuels and renewables are frequently counterposed in public debates about energy futures. These dominant imaginaries mainly revolve around the environmental and economic implications of each. The use of fossil-fuels is linked to nationalism and energy independence, whereas the use of renewables is linked to the climate crisis and sustainable development (Elizondo et al., 2017; Malló, 2021). However, both are focussed on top-down policy frameworks, large-scale technologies and the centralization of energy production and distribution.

In Mexico City, policymakers have used these two imaginaries to create a future vision of the city, where some public spaces and government-owned buildings will be powered by renewables and others by fossil-fuels (SEDECO, 2013). Considering that municipal public markets are officially owned by the city, they were included in such imaginaries of

[1] In this chapter "Imaginaries are landscapes of collective aspirations and/or fears that enact and are enacted by practices and commitments to certain views of the world; also, these encode how life ought (or ought not) to be, and therefore express shared understandings of good and bad" (Sol Cueva, forthcoming). In this sense "imaginaries differ from pure discourse analysis because the former usually focuses on language [...] whereas the latter emphasizes actions and performance with materialization through technology. Imaginaries are not the same as policy agendas or frames either [...], as imaginaries are less explicit and accountable. Nor are they the same as narratives, which are usually extrapolated from past events and serve explanatory or justificatory purposes. Imaginaries instead are instrumental and futuristic, they project visions of what is good, and worth attaining (and also, [...] dystopias worth avoiding" (Sovacool, 2019, p. 174).

the future. Until 2019, it was envisioned that by 2024, municipal public markets would be powered by solar energy (SEDECO et al., 2019). Today, this picture has been erased, keeping markets powered by fossil-fuels and out of the "Solar City".[2]

Arguing that communities outside and beyond centres of power have the capacity to produce and perform other imaginaries (Marquartdt & Delina, 2019), I invited mainly women vendors to creatively and collectively explore the future of four municipal public markets and their energy systems. The activity was a storytelling-game called "Tell me a (un)fortunate story" that used messaging apps as a medium to maintain physical distance during the COVID-19 pandemic. The game consisted of co-creating stories with 16 vendors about the future of municipal public markets and their energy systems.[3] The final versions of the stories were sent to the vendors and discussed with them at the marketplaces.

In this chapter, I present some of the wishes, dreams and fears vendors have while talking about the future and some of their concrete ideas that address energy transitions. I will also elaborate on the energy

[2] *Ciudad Solar* or 'Solar City' is the name that the government is using to talk about Mexico City's energy policy, which focuses mostly on the installation of rooftop PV in the 2019–2024 period.

[3] For more information about the storytelling activity, here I am presenting the game instructions. The instructions were sent as an audio and summarised in a GIF:

AUDIO INSTRUCTION

> You are members of a team of 4 people who will 'Tell us a (un)fortunate story'. The objective of the activity is to create a story collectively about the future of municipal public markets and their energy systems. Each person will be a single element of the story: you can be the main character or characters, the place where the story takes place, the actions that the character performs, or their emotions and moods. I will go from one participant to another and in each turn the participant has to add a bit to the collective story according to her corresponding story element. In total there will be three rounds. I will begin the story by time travelling with you. You will continue creating the story with the help of some questions and sentences. Were the instructions not completely clear? Don't worry, during the activity your doubts will be solved, however, if you still have doubts, do not hesitate to communicate them!
>
> The only rules are that 1) the main character or characters must stay alive and 2) you need to answer to my message with the questions and sentences that same day or the following. I will write down the main ideas and send them to the next participant to read what has been said. [...]. (Sol Cueva, forthcoming).

technologies imagined, as well as issues motivating energy transitions, governance arrangements and the role of different actors in the market's life. However, I will not present these findings as a set of arguments and descriptions that say, "here is what vendors said about A, B or C". Rather, I aim to articulate vendor's alternative visions of the future by sharing a story of a fictional marketplace in 2050. The story will contain some of their concrete ideas and the implications these visions may have in their communities and environments. As such, the story will *show* what markets *can be* if vendors' visions of the future come true (Leavy, 2012), making the story "exploratory, explanatory, hopeful and generative" (Dunlop, 2001 cited in Leavy, 2012, p. 518).

First, the chapter summarizes its feminist and narrative grounds, followed by an explanation of the particular research methods utilized for its development. It then proceeds to tell the story of *La Mercadita*, an imaginary market located in Mexico City in the year 2050. Finally, the chapter ends with a conclusion in which I reflect and push for more "stories for connecting (when the world is falling apart)" (Di Chiro, 2017).

The Beginning: The Grounds

Every story starts from the beginning. This specific one started in the search for hope. Like Giovanna Di Chiro (2017), who was inspired by Rebecca Solnit, I started working with storytelling looking for hope in the tales, histories and stories of people. This is especially important now that hope has become a scarce resource for many of us. Not because of lack of enthusiasm but because wherever we look, capitalism and violence rule, and the possibility of stopping the planet from warming with conventions and treaties has proved disappointing (if not useless).

In storytelling with vendors, I aimed to explore the potential of stories to enrich, deepen and communicate feminist, future and energy studies. My hope was that developing and creating stories with people would provide new imaginaries, visions, ideas, narratives, projects and demands. These would enable the vendors and me to envision a plausible, alternative, non-fossil futures (Adamson, 2016); futures which centre women's histories and lives and tackle gender inequality and environmental injustice. To explore alternative futures, my method of inquiry was feminist standpoint theory, "which examines the experiences of women in order to analyse how they experience and know differently from men, taking into account the intersectionality of gendered experience in its combination with class, race, age and other forms of difference" (Harcourt, 2016, p. 1008). At the same time, I used the notion of socio-technical imaginaries to investigate the landscapes of collective visions of desirable futures that enact and are enacted by commitments to certain views of the world, attainable through and supportive of science and technology (Jasanoff, 2021).

Focussed on creating stories *with* and not *on* Mexican women vendors, I wanted to "challenge notions of expertise and knowledge hierarchies, and [...] to critically examine and challenge representational practices in research and writing" (Harris, 2021, p. 9). Story and storytelling are particularly important for engaging underrepresented communities in research and policies, offering the possibility of shifting the power dynamics of knowledge creation and challenging who listens and who speaks in academia and policy discourse (Ingram et al., 2014).

Moezzi et al. (2017) and Smith et al. (2017) highlight that storytelling can provide data describing phenomena at a variety of levels, including emotional and imaginative. This is particularly relevant in energy studies because "[s]tories offer the possibility of opening up the conversation to participation by people who may not otherwise feel competent in 'energy speak', thus bringing to the fore a wider range of knowledge, insights, and perspectives and a deepened awareness of community values, identities, relationships, cultures, and histories" (Miller et al., 2015, p. 67). Equally, storytelling is important for feminist research since this method "is a way to challenge dominant narratives which erase, oversimplify and universalize women's voices and experiences. It is an unconventional way to explore women's stories and to expand their possibilities as women tell their own stories [,]in their own words" (de Nooijer & Sol Cueva, 2022, p. 238).

By challenging notions of factuality and expertise and working to engage others, storytelling holds the potential to connect individual experiences to broader socio-political realities, explore place-based and everyday practices, and foster collaboration. This is because, when storytelling, participants share imaginations, negotiate meanings and expose elements of the self as well as the broader economic, political and social context. Thus, the experiences reflected in stories are never solely about the individual, but rather about groups and communities, power dynamics, resistances and collaboration (Harris, 2021).

While recognizing these potentials, we must also remain aware of the limits of storytelling and stories. We need to be sensitive to the fact that, when creating stories, people decide what is included and excluded, inevitably silencing other voices and experiences. Therefore, stories need to be understood as "partial truths that offer visions of and insights into situated moments in which they were crafted and about what they were told" (Rice et al., 2020 in de Nooijer & Sol Cueva, 2022, p. 251). Thus, storytelling needs to be performed as an open practice in which narrative closure is resisted.

Bringing together feminist theory and a narrative approach, this chapter stories alternative energy imaginaries explored in the vendors' tales of the future. This enables me to weave together heterogenous experiences into collective knowledges and gives me the chance to creatively present vendors' "understandings of their communities, how they've been treated, what is owed to them by regulators and other dominant groups, and what their future should look like" (Ottinger, 2017, p. 43).

Methodological Disclosure

Feminist theory as well as future studies encourage the use of creative and art-based methodologies in the process of doing research and in presenting its results (Rose, 2013). As such, experimentation with art and imagination has been common in feminist political ecology, environmental justice and nature-society studies (Harris, 2021).

Inspired by such creativity in methods and theory, I adopted a narrative approach and art-based methods in my Ph.D. research as well as in this chapter. During my research, I used a narrative and feminist approach to explore the future of energy with non-hegemonic voices in climate change, energy and future studies. In this chapter, I also chose fictional storytelling as a method to present such desirable futures using the power of fiction, such as its capacity to be a vehicle for greater immersion in what we read and what we retain after it (Leavy, 2018), to promote empathy and collaboration (Leavy, 2012), to resist dominant narratives and to strengthen senses of collective experience and solidarity (Hydén, 2017).

To create the story of an imaginary market called *La Mercadita*, I included elements of the "Fictionalizing Process" explained by Leavy (2018). Leavy suggests using processes of selecting, combining and using self-disclosure to analyse different empirical and story elements, (re)write stories in an orderly way and reveal the texts' "real" and "fictional" nature. First, I selected the main themes and elements of the four stories created by the vendors that represented alternatives to the dominant energy imaginary in Mexico. This was done based on a previous analysis of the four stories, our conversations after them and my observations at the marketplaces. The themes selected were the use of renewable energy technologies in the future, for whom these would be, and how these could be managed. Based on my analysis, I chose dialogue, negotiation and self-organization as effective strategies for the governance of the energy systems and markets.

Second, I combined and reorganized the elements as if the futures imagined in the vendors' stories were real (Leavy, 2018). I consider this as 'a simulacrum' of the world imagined, which means that instead of presenting research results of *how* "reality" *is*, I show *what it can be* (Leavy, 2012). I also made collages with drawings, photographs and illustrations to produce unique visualization and insights into the imagined futures.

Third, I created characters, places and situations that reflected not just the alternative visions imagined by the vendors but also the context in which the stories were created, as well as my personal observations and experiences at the markets. For example, the setting in the following story was developed according to how vendors constructed the markets in their stories, my observations and what has been discussed in the literature.

Fourth, I used parenthetical citations to identify which pieces of this story are directly based on the vendors' interviews and stories. The rest of this story was fictionalized out of my encounters with vendors in 2018, 2019 and 2020, and my long trajectory as a customer of municipal public markets in Mexico City, the city in which I grew up and lived for more than 30 years. In that sense, the characters, places and problematics in the following story "are fictionalized but are being drawn from, and are in response to, lives and living" (Murphy et al., 2017, p. 218).

Lastly, taking into consideration that this is just one story about one possible future, in *La Mercadita*, I did not pretend to describe what will happen in "the future" or to convince you about this vision of the future. Instead, it is expected that readers "[…] try to understand the various developing imaginaries and help to create and disseminate new ones" (Hajer & Versteeg, 2019, p. 132).

The Story (Fig. 11.1)

I couldn't stop thinking. *You are late again Carmen. The clouds are gray, it is going to rain… You forgot to close the dome of the market's milpa*[4] *and your compañeras*[5] *need your fingerprint to close it… Some raindrops might slash down onto the hanging pots, and the crops are not used to the rainwater anymore. The milpa might not survive… Focus, don't stress. You are three minutes away.*

Arriving at the market, I could see the 11 letters above the entrance, "*La Mercadita*", glowing brightly with a purplish spark. Next to it, there were the sleek, vertical touchscreens, for self-ordering in any language, including the 68 indigenous languages spoken in Mexico (Catálogo de Lenguas Indígenas Nacionales; Variantes Lingüísticas de México Con Sus Autodenominaciones y Referencias Geoestadísticas, 2018). The huge

[4] *Milpa* is a traditional intercropping system of regional vegetables and is practised in Mesoamerica. The word milpa comes from *nahuatl*, an indigenous language, and means 'what is sown on top of the plot'. This growing system is configured around polyculture, where crops such as maize, beans, chilli, squash and some herbs are grown together (San Vicente Tello & Jönsson, 2019).

[5] Translated to English, *Compañeras* means companions. In municipal public markets the word can also be used to refer to colleagues who work at the market as well. Among feminists the word is used to refer to those who fight alongside each other against patriarchy and other systems of oppression. In this sense, the word can be equated to the concept comrade. In this text, its meaning is a combination of all the above.

Fig. 11.1 Collage "La Mercadita" by Sol Cueva, 2022

automatic doors opened as I went in, and a cool breeze brushed across my copper-skinned face.

Looking around, all I could see was a crowd of people shouting and shoving. Apart from our regular visitors, there were holograms and people with spandex suits walking through the corridors that divide and organize the market in a nearly perfect grid (Repoll, 2010). They were taking photos as fast as they could, interrupting the ones who were purchasing items and bartering. The actions of these unusual visitors were almost robotic, smiling, picking up the item, taking a photo and always applauding after a vendor shouted "*¿Qué va a llevar marchanta*[6] ?"[7]

Walking through the main hall, I could smell papayas and oranges, which tempted me to slow my pace. Instead, I said good morning and rushed, returning to my mission. I crossed the communal kitchen and ran into the shared utility room where I finally reached the controls to close the dome.

Once the *milpa* was protected, I took a step backward and focussed on the large group that was walking towards one of the stalls. It was a vegetable stall, in which a *mestiza*[8] woman like me was packing a bunch of limes under the lamp light, asking questions about kilos and products. I noticed a peculiar tone of voice coming from the centre of the group. There he was, Professor Carcar Solar,[9] the most famous influencer-scholar, giving a live presentation of *La Mercadita*.

"Here you can see a vendor with her customer. Can you see how the vendor is selecting limes and giving them to this woman? Yes, you guessed right, here they don't use food-walls, but they have products at display. People can pay with credit cards and biometric systems, but they seem to prefer social interaction, touching the products and smelling them,

[6] *Marchanta* is a word of popular use by vendors in Mexican markets used to address the buyers. The diminutive is *marchantita* and male *marchantito*. Possibly, the French introduced this word in Mexico, as it is similar to *marchand* or merchant.

[7] Translated to English, the sentence means "What are you gonna buy, *marchanta*?"

[8] *Mestiza*, as understood in Mexico, generally refers to a woman who has been the result of racial and cultural mixture between Europeans, Indigenous and/or African descendants (López-Beltrán & García Deister, 2013).

[9] This character was created by me, it did not appear in the stories or the interviews. It was thought of as a critique of some academics' behaviour, who approach the world as a laboratory and participants as 'objects' to study, and who try to capitalise on people's knowledge, efforts and experiences.

funny right!" Carcar mentioned. "Why did you decide to keep the folklore of your little stall?" he asked the vendor, and when she was about to respond, the professor interrupted "Well, actually…" and then he ignored my *compañera*, sharing his own interpretation. He did the same to the vendor in the next stall and the next.

As one of the vendor's representatives, I could not tolerate his attitude. I made my way through the hologram devices and the people gathered to ask Carcar about his presence in our space. "*Oiga señor*, if you are going to talk about *La Mercadita,* I suggest you engage with us, the vendors and customers. We built this space, and we keep it running every day. We know its history, because we are part of it." But Car, as he likes to be called, was not interested in our story, he was just trying to sell his tours of what he calls "the city's food scene" (Wattenbarger, 2019). He responded, using his instant translation device to speak in "common people's language", with an explanation of how social media works and how the experience of a woman like me was not fascinating enough for his audience.

I was crestfallen and furious at the same time. Many questions popped into my mind. *What should I do now? How dare he treat us like this? We are not lesser than him, and more importantly, it is our story to tell, not his.*

I could feel myself filling with anger. I looked him straight in the eyes and took a deep breath, remembering all the time that me and my *compañeras* needed to stand up and demand better treatment and a better market. Then, I said to the crowd, almost shouting "Just so you know what *La Mercadita* is, it seems fair that I tell you our story. It is the story of how this marketplace came to be. And it is the story of the lessons we learned along the way. In our journey, we were not alone. We were imagining multiple futures for our marketplaces with and for vendors…"

Just like that, I was not mad anymore. I saw the other vendors nodding and cheering, the people in the crowd were listening. I felt powerful. So, I continued (Fig. 11.2).

"To be honest, when we started to imagine and to make *La Mercadita* a reality, we thought it was like an oasis. It was a fertile idea in an arid region. Dreaming of its future existence, the vision of a strengthened community and market, gave us relief. At the same time, it made us believe that we were losing our minds in hallucinations."

Fig. 11.2 Collage "Our Story" by Sol Cueva, 2022

"In 2025, in the middle of the chaos and confusion that the COVID-19 pandemic left behind, many vendors tried to avoid the loss of more lives and livelihoods (Castellanos, 2021a, 2021b). Numerous markets started to explore ways to resist what seemed to be a fatal destiny. Stalls were closing, markets were burned down (Corona, 2021), and supermarkets were built on top of their ruins. Our markets were being buried, just like the great pyramids of this territory called Mexico were buried by the Spaniards hundreds of years ago. On top of this, we were experiencing complete power outages. We knew this was coming when in 2021, from one moment to the other, large parts of Mexico City went dark ('Mexico Suffers Another Day of Rolling Blackouts Due to Storm', 2021). You might imagine how stressful it was, we were not able to work for hours until the electricity came back.

"We had already known for decades that structural renovations were needed at the markets and that we were not able to afford them. For years we, the vendors, absorbed the costs of small renovations here and there, but we did not have the means nor the capacity to make major changes (Meneses Reyes, 2011). In particular, the energy, water and sanitation systems were as old as the markets themselves; 70–75 years-old in 2025, and the resources to transform them were out of our reach (Liliana, personal communication, 26 November 2020). Plus, we were not able to decide how and for whom energy would be produced or how to deal with droughts and floods in Mexico City, two sides of the same coin. We felt like our hands were tied!"

"It is true that there were policies and programs in place since the beginning of the century. We would hear politicians and entrepreneurs saying 'this policy will rescue urban heritage' and 'this project will modernize food markets' (Delgadillo, 2018). However, these were insufficient and inadequate to face the challenges and bring the solutions that our communities were demanding (Giglia, 2018). There was always, either a lack of money, or an invisible hand that favoured supermarkets and malls instead of the municipal public markets" (Delgadillo, 2018).

Then, a hologram woman said "Yes, I remember campaigns promoting the modernization of markets. Close to my neighbourhood, a market was renovated. It was super cute after. We could find hip restaurants and cafés. The only weird thing was that all the vendors were new. It was like the renovation replaced the vendors who worked at the market before. But *La Mercadita* doesn't look like that. What did you do here?"

"Well, the first thing we did was to create a commission. It was our goal to travel around the city to collect ideas and to learn from other market's experiences, like merchant's caravans moving across continents to find valuable goods. And so, we started the journey that same week. It was March 2025 and we began to visit the remaining municipal public markets in Mexico City, to learn from and with them (Fig. 11.3)."

"We travelled for almost a year." Said Lulú, who took the floor, continuing with the story. She was passing by, going from the communal kitchen to the care centre where she would pick up her granddaughter. Lulú was also a vendor and part of the caravan. So, I invited her to share her experience.

"Carmen, the other vendors and I visited the markets to pay close attention to the challenges they were facing and the ways they were solving them. It was amazing! Some markets were moving away from processed food and the use of electronic and digital technologies. They went back to selling only fresh fruits, vegetables and seeds and were using human-powered machines. Other markets were installing solar panels, buying waste-to-energy systems, and using robotic cleaners." She continued, "For example, in Mercado Rosa Torres, we were welcomed by Lucky, the robot-girl cleaner, who was about to clean the market's corridors, as she does every morning and night. She is powered by the waste-to-energy system for which she collects and recycles waste (Liliana et al., forthcoming). In Mercado San José, they were using banana peels to power their information screens and biometric payment systems... and they were even planning to use a micro-hydropower generator and install small wind turbines (Gabriela et al., forthcoming). In Mercado San Joaquín, they were expanding the solar panel system to all the market, inspired by a woman butcher who turns a regular stamp-clock into a solar-stamp-clock (Erika and Jess, forthcoming). Literally, some markets were covering all the renewable options!" Lulú exclaimed.

Then I added "I also remember that in markets like Mercado San Mateo Tlaltenango, vendors managed to redesign the stalls and vertically expand them, in combination with the installation of individual solar panels and rainwater harvesting systems for the entire market. They

Fig. 11.3 Collage "The Caravan of Wonders" by Sol Cueva, 2022

even had a vegetable garden for the *cocinas corridas*[10] (Claudia et al., forthcoming)".

Yes! And when we asked the vendors why they decided on all those transformations, the response was similar in every market. Vendors didn't want to struggle with water or energy scarcity, nor pay high energy prices, and they wanted to care for the environment and their livelihoods. Vendors wanted to keep and recover some traditional practices while being modern at the same time (Claudia et al., forthcoming). By this, I mean markets wanted the newest technologies, but not at the cost of their practices and wellbeing (Liliana et al., forthcoming), Lulú pointed out.

"It all makes sense!" someone from the audience raised her hand and said, "Municipal public markets wanted new paint on the walls, but they also needed reliable services, and to keep running without turning into supermarkets."

"That was exactly our conclusion after visiting the markets. Well noted! But then again, that was not all!" Lulú replied. "It was time for us to present these ideas to all the vendors, after which we needed to come to an agreement. It was time for us to re-shape our market and to re-think services, products, and ways of interacting with each other and our communities. Carmen, would you tell them about this? I must go" Lulú said, giving me a hug. As she walked towards the care centre, she shouted "Do not forget to visit the kitchen, I have made fresh zucchini blossoms in *salsa verde*[11] with ingredients from our *milpa*. I promise that you will love them. Bye (Fig. 11.4)."

Oh, me oh my, I have to tell people about the transformations we made here and how we did it. Not an easy task. Let me think... "Everybody, I think that the best way to continue the story would be by showing you what we—"

"Wait a minute sweetheart." Carcar stepped forward and faced the audience. "I think I can take over from here. I have very nice holographics to show the transformations."

[10] *Cocina corrida* is a business that sells meals of several courses at a fixed price, eaten about 2 and 4 pm. It is called *corrida* (run) because it is expected that you eat this meal in a maximum of 30–50 minutes.

[11] *Salsa verde* is a spicy sauce made of tomatillos, onion, green chili and cilantro.

Fig. 11.4 Collage "New Endeavours" by Sol Cueva, 2022

"No thank you, sweetheart. I can do it. Actually, I was about to give some examples of what we have done here and explain why. Is this okay for everyone?" and the group nodded "Great, follow me."

I walked calmly, leading the visitors from the communal area to one of the four main corridors. This way people could take some time to look around the marketplace. The corridors were wide, allocating separated stalls on either side of them (Repoll, 2010). The stalls were not grouped according to their commercial activities and products as it was years ago (Meneses Reyes, 2011). In *La Mercadita*, vegetables, fruits, repair shops, etcetera, were scattered throughout the market.

"Here we are!" I stopped in front of my stall. It was a metal structure of two floors, shear walls and sliding glass doors on the ground floor. A 3D printer, tools and different materials such as fabric, wood and paper were visible through the glass doors. There was a table at the middle of the room, and there were shelves displaying shoes, cutlery, pencils, stencils and toys all the way to the back.[12] "Ten years ago, this stall used to sell dried seeds and *mole*,[13] as you can see in the photo hanging next to the door. Today it is a 3D workshop. 'Why did Carmen change her stall?' You might ask. Well, my *compañeras* and I decided to keep just one stall with dried products in the market and to set up a workshop instead. This way, we would avoid competition among us and serve diverse local needs (Brenda, personal communication, 25 November 2020). Let me show you a pair of shoes we have been working on."

I picked up and showed a pair of shoes with woven soles of natural fibre, which were having a major revival after being out of fashion for nearly 50 years. Suddenly, a written message popped-up in Car's holographic videocall chat asking, "Do you have other services like this?"

[12] The idea of a communal workshop was completely imagined by me. It was inspired by spaces such as 'Hackerspace Rancho Electrónico' (collective project of co-construction of knowledge), 'Casa Gomorra' (collective project of dissent, bodies, pleasures and politics) and 'Enchúlame la Bici' (communitarian workshop to build and repair bikes), among others in Mexico City.

[13] *Mole* is a traditional marinade and sauce in Mexican cuisine. Generally, *mole* contains dried chilies, nuts, chocolate and spices such as black pepper, cinnamon or cumin. They can range from bittersweet to spicy.

"Yes, at *La Mercadita* we have more services that share the same principles. We have my stall which is a workshop, the kitchen, the *milpa* and the care centre. The four of them serve a purpose for the market and its community, they are led by us, and we assume mutual responsibility towards them. For example, the kitchen is where vendors and visitors can go to drink, eat, bake or even to socialise. It is run by the workers of the old *cocinas corridas*. So, the team cooperates in collecting the ingredients, cooking, and cleaning the kitchen, in coordination with the rest of the market which also collaborates with them in clear tasks.[14] Actually, if you look over there you can see that two people are working in the *milpa* right now" I pointed to the woman and man standing beneath the maize leaves hanging from pots, carrying large hand-woven baskets. "These vendors are harvesting chilies and squash that will be used for dishes in the kitchen and sold at the market later (Claudia et al., forthcoming; Gabriela et al., forthcoming). Finally, in the care centre we provide and receive self-managed care services such as basic healthcare and childcare. I could go on and on about the care centre, but the truth is that this project has been set up recently. So, we are still experimenting and learning from it. I hope I can tell you more next time you visit us. For now, I will tell you that these services were recovered from the old markets' designs, in which there were stalls but also libraries, day-care centres, among others (Delgadillo, 2016; Meneses Reyes, 2011; Repoll, 2010). Nice, right?" And people reacted by having their watches pop up a hologram GIF of a puppy holding a sign that said "Mind blowing" on it.

"Talking about experimentation, it is important to point something out. Seeing the market today, you might think that we knew exactly what we wanted and how to achieve it, when in reality, things were and are not so straightforward" I added. "The processes of thinking together

[14] The idea of four communal services was mostly produced by me. It was inspired by San Mateo's and San José's stories in which community tasks are presented. It was also based on the way vendors organise market clean-up days, block parties, and planning meetings, as well as real-world examples of collective work such as *tequio* in Oaxaca and other parts of Mexico. *Tequio* is a form of organisation of labour in which all healthy members of a community must participate with the same regularity in equally arduous community activities (Zolla & Zolla Márquez, 2004).

and putting things in place took years of trial and error, of testing and piloting projects (Erika, personal communication, 27 November 2020). This did not happen from one day to the other. Little by little, market by market, we were gathering puzzle pieces and learning how to fit them together, in a way that made sense for our community and every stall. I mean, we were not planning to copy-paste the ideas of other markets. Why would we pretend to be exactly the same? For instance, we first thought about having a marketplace full of the latest technology, focussing just on e-commerce. But we remembered what happened in Mercado Rosa Torres: transforming the market in that way caused disunity between vendors, distance from the community and dissatisfied customers. 'Customers love to inspect their avocados before buying them' (Liliana et al., forthcoming), I remember a vendor saying to us. So, after long hours of presenting the technologies and projects, after discussing their cons and pros, and making their benefits for the majority visible, we came to an agreement (Claudia et al., forthcoming; Gabriela et al., forthcoming; Liliana et al., forthcoming). We decided to renovate the market, introducing new payment methods, screens, and new appliances, but maintaining direct contact with customers and products at display" (Claudia et al., forthcoming; Erika & Jess, forthcoming; Gabriela et al., forthcoming; Liliana et al., forthcoming). Even the most sceptical vendors agreed.

"We also considered other things that were shared by the markets we visited. In Mercado San José, vendors told us that they were using so many emerging technologies that the high energy demand did not fit their energy system. Thus, they needed to install a complex hybrid renewable energy system to meet their energy needs (Gabriela et al., forthcoming). When we asked ourselves if we could run *La Mercadita* on energy technologies such as hydroelectric dams, wind turbines and similar, we said no."

I glanced at the audience, who were looking at me with confused expressions. After many years of pushing states to support the goal of 100% renewable energy technologies, what I just shared sounded completely contradictory to them. I took a deep breath. "Please, don't look at me like that. Of course, we also wanted to get rid of fossil-fuels, protect the environment, and have energy free of charge or at low cost! But, at the same time we wondered which type of energy systems we needed, for what and for whom. We wanted to make a decision that was centred on us, the complete market and every stall, not just on energy technologies. So, we gathered in vendor's assemblies to decide what we wanted" (Claudia et al., forthcoming; Gabriela et al., forthcoming).

"After discussing the options, in several general assemblies, we decided that we needed a mix of small-scale renewable energy technologies. Also, we were interested in taking advantage of architectural adjustments that maximise the benefits of the heat and light gained from the sun moving and the wind blowing. For example, look at the lamps above us. They look like ordinary electric lamps, right? Well, they are not. These are pipes of reflective materials that capture and bounce back the sunlight (Mayhoub, 2014)."

"So, these lamps don't use any electricity!?!?" Car exclaimed, incredulous.

"Nope, they don't. Now look in between the lamps. Do you see the small, spinning turbine? It is a wind driven vent that continuously replaces trapped, stale air with cleaner, cool air from outside. It does not use electricity either and it is a technology that has existed at the markets since the last century. And…I am sure that you also saw the solar panels on our roof when entering the market. Similarly, we installed a rainwater harvesting system to better manage our water needs (Claudia et al., forthcoming; Gabriela et al., forthcoming). Also, we have a micro waste-to-energy system to manage waste and generate power. You can't see these last two, because they are in the closed loading and unloading area. My point is that what we decided was "to combine modern and natural energy' (Claudia et al., forthcoming) to meet our needs.

"Once we knew which type of energy we wanted, we agreed that energy would be produced and controlled by us, therefore we would not be connected to the cables on the streets (Claudia et al., forthcoming; Erika & Jess, forthcoming; Gabriela et al., forthcoming; Liliana et al., forthcoming). Finally, we made alliances with the government and experts, considering that they have more money and information than us" (Mario, personal communication, 26 November 2020).

"Aha, I knew it. You couldn't do this by yourself. I mean, we see how the energy industry does its best to give us energy, and they can't manage." Car voiced.

"Just like other markets, we learned that alliances were necessary to achieve our goals. The majority didn't want a third party to control and produce energy and agreed that we couldn't do everything by ourselves with little resources. Therefore, WE achieved this, but with the support of others. We wanted to collaborate with people and institutions that supported our journey and who acknowledged that energy needs and interests are only known by us, the vendors (Jess, personal communication, 27 November 2020). We did not and do not want any more lies, empty words or random ideas. With this, I mean that there is no room for those who want to replace the marketplaces with supermarkets or gourmet markets, neither is there room for alliances that want to renew the marketplaces but displace the vendors or alliances that don't recognize, revalue, improve and protect our markets" (Delgadillo, 2020, p. 10).

"So, I would kindly ask you, Carcar, to leave *La Mercadita*. I have had enough of your behaviour which, sadly, inhibits OUR collective ways to flourish." I said while opening up my arms, as if I was wrapping the audience and the vendors in a hug, making the "we" gesture.

"For the rest of you, please stay as long as you want. I hope you enjoyed our story and that you will be back soon to see what other things we are creating here, or even better, to join us in making *La Mercadita* into what we have dreamt. Have a wonderful rest of the day!" I said smiling, before heading towards the flourishing veggies under the dome. It was time to go back to work (Fig. 11.5).

Conclusions

In an attempt to explore the future visions of municipal public markets in Mexico City, I used storytelling, a non-traditional method in energy research. In this chapter I wove together "other ways of doing knowledge", inviting readers to think about alternative ways of imagining energy futures.

In order to pluralize the voices of these futures and to challenge the "male technical" voice in energy transition conversations, my text tapped into a range of historically underrepresented voices, centrally those of women vendors. By using an imaginary setting, informed by the visions of vendors, the text gave insight into the role of energy in women's daily activities and how changing the socio-technical energy systems might impact their community, their day-to-day lives and the city in general. It showed the possible societal, technological, environmental and political influences on the created story world.

For example, based on the stories and the conversations with vendors, I created strong female characters who work at markets and collaborate with other vendors to make markets better places in the future. During

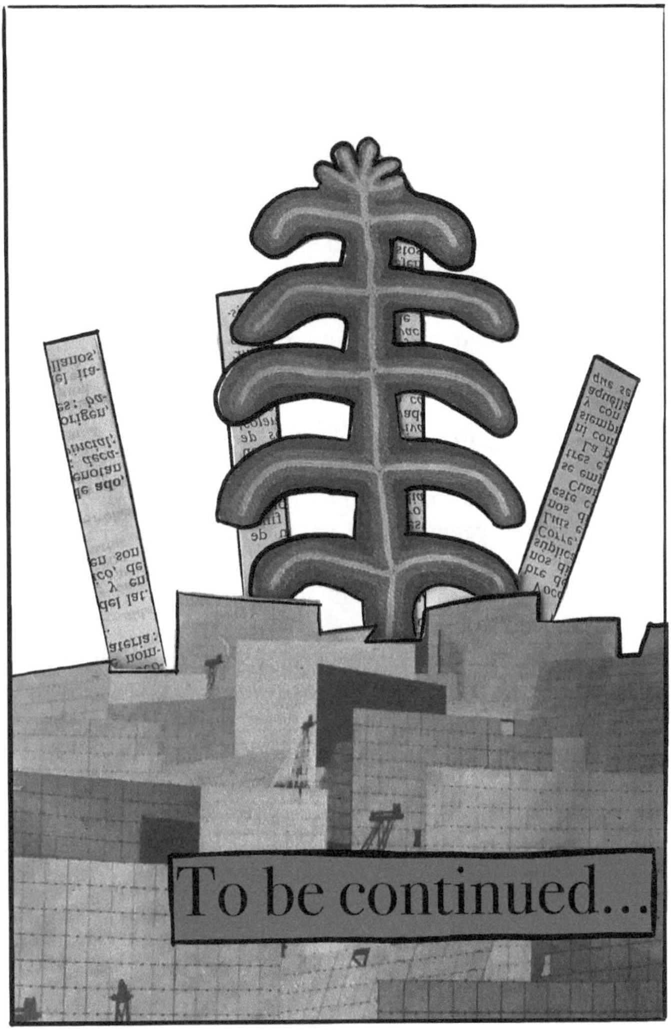

Fig. 11.5 Collage "To be continued..." by Sol Cueva, 2022

the fictionalizing process, I used the characteristics of the marketplaces and their energy systems narrated by these women vendors. Vendors described new energy technologies and fuels, such as solar energy and waste-to-energy technologies to face future energy challenges. Also, vendors described governing structures in which they install and control renewables and forums in which all vendors participate. In this sense, storytelling facilitated complex reflections about the energy futures.

The created story demonstrates that the way people understand and engage with energy systems is informed by the cultural and social as well as the technical context. The chapter presents an image of what a world without fossil-fuels would look like in a specific place and time, and how methods such as storytelling and speculative fiction can contribute to research.

Finally, in searching for hope, I found hopeful visions of the future. During the storytelling activity, the vendors had the option of telling a fortunate or unfortunate story, and all of them chose the first. This did not mean that vendors cannot shine a spotlight on injustices and inequalities, power relations and oppressions—they do—but they decided to focus on a future without major conflicts and disruptions. What the stories showed us is that hope can bring people together and form a basis for collective action. Vendors imagined better futures for themselves, others like them and for people who are willing to contribute to the flourishing of their communities.

Hopefully, this exercise will invite others to keep exploring non-hegemonic imaginaries in feminist political ecology and energy studies and provide some ideas on how collectively imagining energy futures can inform energy, future and feminist studies.

Funding: Lillian's research was supported by the Secretaría de Energía and the Consejo Nacional de Ciencia y Tecnología (SENER-Conacyt) through its post-graduate grant programme. In addition, this chapter was funded by the Wellbeing Ecology Gender and cOmmunities Innovation Training Network (WEGO-ITN) funded by the European Union's Horizon 2020 research and innovation programme under the Marie Sklodowska-Curie grant agreement No. 764908-WEGO 2018-2021.

References

Adamson, J. (2016). Humanities. In J. Adamson, W. A., Gleason, & D. N. Pellow (Eds.), *Keywords for environmental studies* (Print, pp. 135–139). New York University Press.

Birch, K. (2017). Techno-economic assumptions. *Science as Culture, 26*(4), 433–444.

Brenda. (2020, November 25). *Interview San José* [Personal].

Castellanos, Z. (Host). (2021a, February). *Iniciemos el viaje: Sistema de Abasto Tradicional y su importancia en la vida de la Ciudad Chilanga* (No. 1) [Audio podcast episode]. In *OrigenEs.* Retrieved 13 February 2021a, from https://open.spotify.com/episode/2ICxTdq2pvzBROMj5SpcrO?si=zEV9G3yLTwWml-pWWlP7dw

Castellanos, Z. (Host). (2021b, February). *Mujeres en el comercio en tiempos de Pandemia* (No. 6) [Audio podcast episode]. In *OrigenEs.* Retrieved 22 April 2021b, from https://open.spotify.com/episode/6VrzL3RlXeafvV0QDl3r0x?si=mL_Mn7LRRFCwd5014KZaGA

Catálogo de Lenguas Indígenas Nacionales; Variantes Lingüísticas de México con sus autodenominaciones y referencias geoestadísticas, Diario Oficial de la Federación, Mexico § Primera (2018). https://www.inali.gob.mx/pdf/CLIN_completo.pdf

Claudia, Marisela, Color, & Fernando. (forthcoming). *Mi Mercado San Mateo Tlaltenango 2050* (L. Sol Cueva, Trans.) .

Corona, S. (2021, November 4). ¿Cuántos mercados se han incendiado en la CDMX? *El Universal online.* https://www.eluniversal.com.mx/metropoli/mercado-sonora-cuantos-mercados-se-han-incendiado-en-la-cdmx

de Nooijer, R., & Sol Cueva, L. (2022). Feminist storytellers imagining new stories to tell. In W. Harcourt, K. van den Berg, C. Dupuis, & J. Gaybor (Eds.), *Feminist methodologies: Experiments, collaborations and reflections* (pp. 237–255). Palgrave Macmillan. https://doi.org/10.1007/978-3-030-82654-3_11

Delgadillo, V. (2016). La disputa por los mercados de La Merced. *Alteridades, 26*, 57–69.

Delgadillo, V. (2018). Markets of La Merced: New frontiers of gentrification in the historic centre of Mexico City. In S. González (Ed.), *Contested markets contested cities. Gentrification and urban justice in retail spaces* (pp. 19–35). Routledge. http://contested-cities.net/wp-content/uploads/2018/08/Chapter-2.pdf

Delgadillo, V. (2020). *Mercados Vivos* [Contest]. Concurso Mercados Post Covid-19. https://www.academia.edu/44152385/MERCADOS_VIVOS

Di Chiro, G. (2017). Seaweed, "soul"-ar panels and other entanglements. In J. Adamson & M. Davis (Eds.), *Humanities for the environment: Integrating knowledge, forging new constellations of practice* (Print, pp. 70–87). Routledge.

Elizondo, A., Perez-Cirera, V., Strapasson, A., Fernandez, J. C., & Cruz-Cano, D. (2017). Mexico's low carbon futures: An integrated assessment for energy planning and climate change mitigation by 2050. *Futures, 93*, 14–26. https://doi.org/10.1016/j.futures.2017.08.003

Erika. (2020, November 27). *Interview San Joaquín* [Personal].

Erika, & Jess. (forthcoming). *Mi mercado San Joaquín- Zona 22 2050* (L. Sol Cueva, Trans.) .

Gabriela, Papelería Moce, Odilón, & Brenda. (forthcoming). *Mi Mercado San José 2050* (L. Sol Cueva, Trans.).

Giglia, A. (2018). *Comercio, Consumo y Cultura en los mercados públicos de la Ciudad de México.* Universidad Autónoma Metropolitana, Unidad Iztapalapa, División de Ciencias Sociales y Humanidades, Departamento de Antropología. http://dcsh.izt.uam.mx/departamentos/antropologia/wp-content/uploads/2020/01/libro_comercio_cultura_dra_giglia.pdf

Hajer, M., & Versteeg, W. (2019). Imagining the post-fossil city: Why is it so difficult to think of the new possible worlds? *Territory, Politics, Governance, 7*, 122–134.

Harcourt, W. (2016). Gender and sustainable livelihoods: Linking gendered experiences of environment, community and self. *Journal of Agriculture, Food, and Human Values, 33*(4), 1007–1019. https://doi.org/10.1007/s10460-016-9757-5

Harris, L. M. (2021). Towards enriched narrative political ecologies. *Environment and Planning E: Nature and Space, 5*(2), 1–26. https://doi.org/10.1177/25148486211010677

Hydén, L. C. (2017). Entangled narratives: Collaborative storytelling and the re-imagining of dementia. *Oxford University Press*. https://doi.org/10.1093/oso/9780199391578.001.0001

Ingram, M., Ingram, H., & Lejano, R. (2014). What's the story? Creating and sustaining environmental networks. *Environmental Politics, 23*(6), 984–1002. https://doi.org/10.1080/09644016.2014.919717

Jasanoff, S. (2021, February 25). *Renewable futures and the future of renewables* [Keynote]. The Leibniz Research Alliance on Energy Transitions Energy Futures Conference, Online. https://www.leibniz-energiewende.de/konferenzen/2021-energy-futures-emerging-pathways-in-an-uncertain-world/conference-recordings

Jess. (2020, November 27). *Entrevista San Joaquín* [Personal].

Leavy, P. (2012). Fiction and the feminist academic novel. *Qualitative Inquiry, 18*(6), 516–522. https://doi.org/10.1177/1077800412442813

Leavy, P. (2018). Fiction-based research. In *Handbook of arts-based research*. The Guilford Press. https://search-ebscohost-com.eur.idm.oclc.org/login.aspx?direct=true&db=nlebk&AN=1535661&site=ehost-live&scope=site

Liliana. (2020, November 26). *Interview Rosa Torres* [Personal].

Liliana, Mario, Silvia, & Marisela. (forthcoming). *Mi Mercado Rosa Torres 2050* (L. Sol Cueva, Trans.).

López-Beltrán, C., & García Deister, V. (2013). Scientific approaches to the Mexican mestizo. *Ciências, Saúde-Manguinhos, 20*, 391–410.

Malló, O. (2021, October 16). Petronacionalismo, antiextractivismo y capitalismo verde en el México actual. *La Jornada del campo.* https://www.jornada.com.mx/2021/10/16/delcampo/articulos/petronacionalismo-antiextractivismo.html

Mario. (2020, November 26). *Interview Rosa Torres* [Personal].

Marquartdt, J., & Delina, L. L. (2019). Reimagining energy futures: Contributions from community sustainable energy transitions in Thailand and the Philippines. *Energy Research & Social Science, 49*, 91–102.

Mayhoub, M. S. (2014). Innovative daylighting systems' challenges: A critical study. *Energy & Buildings, 80*, 394–405. https://doi.org/10.1016/j.enbuild.2014.04.019

Meneses Reyes, R. (2011). *Legalidades públicas: El derecho, el ambulantaje y las calles en el centro de la Ciudad de México (1930–2010).* Universidad Nacional Autónoma de México y el Centro de Investigación y Docencia Económica. http://ru.juridicas.unam.mx/xmlui/handle/123456789/11968

Mexico suffers another day of rolling blackouts due to storm. (2021, February 16). *The San Diego Union Tribune.* https://www.sandiegouniontribune.com/business/nation/story/2021-02-16/mexico-suffers-another-day-of-rolling-blackouts-due-to-storm

Miller, C. A., O'Leary, J., Graffy, E., Stechel, E. B., & Dirks, G. (2015). Narrative futures and the governance of energy transitions. *Futures, 70*, 65–74.

Moezzi, M., Janda, K. B., & Rotmann, S. (2017). Using stories, narratives, and storytelling in energy and climate change research. *Energy Research & Social Science, 31*, 1–10. https://doi.org/10.1016/j.erss.2017.06.034

Murphy, M. S., Huber, J., Clandinin, D. J., Caine, V., Steeves, P., & Estefan, A. (2017). Exploring the purposes of fictionalization in narrative inquiry. *Qualitative Inquiry, 23*(3), 215–221. https://doi.org/10.1177/1077800416643997

Ottinger, G. (2017). Making sense of citizen science: Stories as a hermeneutic resource. *Energy Research & Social Science, 31*, 41–49. https://doi.org/10.1016/j.erss.2017.06.014

Repoll, J. (2010). Consumo y Usos de la Televisión en los mercados públicos de la Ciudad de México. *Comunicación y Sociedad, 14*, 83–108.

Rose, D. B. (2013). Slowly ~ writing into the Anthropocene. *TEXT*, 1–14. https://doi.org/10.52086/001c.28826

San Vicente Tello, A., & Jönsson, M. (2019). Landrace maize diversity in milpa: A socio-ecological production landscape in Soteapan, Santa Marta Mountains, Veracruz, Mexico. *Satoyama Initiative*. https://satoyama-initiative.org/case_studies/landrace-maize-diversity-in-milpa-a-socio-ecological-production-landscape-in-soteapan-santa-marta-mountains-veracruz-mexico/

SEDECO. (2013). *Política de Protección y Fomento para los Mercados Públicos de la Ciudad de México (2013–2018)*. Secretaría de Desarrollo Econónimo de la Ciudad de México (SEDECO). http://www.sedecodf.gob.mx/archivos/Politica_de_Proteccion_Mercados_Publicos.pdf

SEDECO, Energética, & D.G.d.D.y.S. (2019). *Política Energética Sustentable para la Ciudad de México*. Secretaría de Desarrollo Econónimo de la Ciudad de México (SEDECO). https://www.sedeco.cdmx.gob.mx/programas/programa/CDMX2019

Smith, J., Butler, J. R., Day, R. J., Goodbody, A. H., Llewellyn, D. H., Rohse, M., Smith, B. T., Tyszczuk, R. A., Udall, J., & Whyte, N. M. (2017). Gathering around stories: Interdisciplinary experiments in support of energy system transitions. *Energy Research & Social Science, 31*, 284–294.

Sol Cueva, L. (forthcoming). *Women's visions: Futuring energy for Mexico City's Public Markets* (PhD). International Institute of Social Studies-Erasmus University Rotterdam.

Sovacool, B. K. (2019). *Visions of energy futures: Imagining and innovating low-carbon transitions*. Routledge, an imprint of the Taylor & Francis Group. https://doi.org/10.4324/9780367135171

Wattenbarger, M. (2019, December 9). Are we killing the markets of Mexico City? [Eater]. *Eater*. https://www.eater.com/2019/12/9/20963654/mexico-city-public-markets-in-trouble-over-tourism-gentrification

Zolla, C., & Zolla Márquez, E. (2004). 24.- ¿Qué son el tequio, la gozona, la faena, la fajina, el tequil, la guelaguetza, el trabajo de en medio y la mano vuelta? In *Los pueblos indígenas de México, 100 preguntas* (Second). Universidad Nacional Autónoma de México; E-Book. https://www.nacionmulticultural.unam.mx/100preguntas/pregunta.php?num_pre=24

CHAPTER 12

The Territory of Our Body: A Conversation on Urban Environments in the Andes and Their Bodies

Agustina Solera and Mariana Jesús Ortecho

Introduction

The reflections in this chapter are inspired by our concern that "knowledge" needs to be rethought in order to face the current socio-environmental risks to which Western society is exposed. In our dialogue, we reflect on what can be learned from the Andean worldview, a Latin American critical theoretical perspective and decolonial thinking.

A. Solera (✉)
International Institute of Social Studies, Erasmus University Rotterdam, The Hague, The Netherlands
e-mail: solera@iss.nl

M. J. Ortecho
Centro de Investigaciones y Estudios sobre Cultura y Sociedad (CIECS), Consejo Nacional de Investigaciones Científicas y Técnicas (CONICET) y Universidad Nacional de Córdoba (UNC), Córdoba, Argentina
e-mail: marianaortecho@artes.unc.edu.ar

W. Harcourt et al. (eds.), *Contours of Feminist Political Ecology*, Gender, Development and Social Change,
https://doi.org/10.1007/978-3-031-20928-4_12

This conversation outlines a journey that starts from the consideration of the "urban" as a modern principle of the organisation of social space capable of expressing profound ethical-aesthetic guidelines that collide with South American rooted ways of inhabiting.

As we suggest throughout the chapter, other-than-modern ways of settling community or settling community "otherwise" (Escobar, 2007), in this case within the framework of the Andean cosmovision, are based, fundamentally, on an understanding and sense of the relationship between the domains of the social and the natural that is different from the categories of Western thought. The hierarchical relationship characteristic of the Western matrix of knowledge, that reduces nature to a resource, is irreconcilable with the Andean conception of nature and social relations, as in this worldview no form of life can be instrumentally valued.

These marks of the Andean worldview—explored in our text through a series of stories—constitute a cultural heritage that "teaches", "shows" a particular way of living that we believe provides key elements for possible answers to our current social and environmental crisis.

Reaching this cosmovision implies that the reader is disposed to "let oneself be taught"; that is, they possess the willingness to move away from the relational position of domination over everything that one wants to know that is characteristic of the modern/colonial epistemological matrix.

Thus, listening to the Andean worldview requires us to be open to new meanings of such fundamental concepts such as "body", "territory", or even "sidereal space"—since looking at a different perspective in relation to "ways of living" implies in the first place a deep ontological reconsideration.

In the framework of Western culture, the notion of the body, for example, is restricted to a physical and individual dimension. This conception is irreconcilable with the Andean one that understands the body as a manifestation that exceeds the physical and material dimension and is not restricted to the domain of the individual but is in a complex connection with others. This consideration defines notions of "body", "territory" and hence any other spatial dimension as clearly different and distant from the same terms understood within the framework of the modern/colonial matrix.

By encountering these matters, we do not intend to expose and settle differences between a dominant Western knowledge and an Andean-silenced one, but to rehearse contributions coming from traditions and memories otherwise that could enrich possible answers to the needs of our particular civilisational moment.

With our reflections, we long to contribute to a critical process of cultural transformation and at the same time to show that the traits that are mentioned here, either at the habitational level or a semiotic-expressive level, come from a deep dimension that must be understood from a perceptual plane that exceeds the possibilities of rational evocation.

The search to reformulate the idea of "knowledge" by decolonial thinking, that is, to endow new meanings based on the plurality of perspectives in the world, has emerged in recent decades from different disciplinary trends and areas of knowledge (like Anthropology, Psychology, Sociology, Discourse Studies, Gender Studies, Feminist Theory, and so on), as an epistemic response to the Enlightenment-based Western epistemology characterised by a rational, abstract, universal, decontextualised and disembodied knowledge. This "(re)territorialization of knowledge" has opened a fertile space for contributions to knowledge from other-than-Western gnoseological traditions based on a long-lasting experience—like Andean gnoseology, the subject of our chapter.

Many of the elements promoted and developed by decolonial thinking (DT) encounter and connect to the ones promoted and developed by feminist political ecology (FPE). At the same time, both perspectives can be seen to converge with elements evoked from the Andean worldview. Our reflection embraces and tries to build connections across these two critical and politically engaged positions.

Both DT and FPE are deeply concerned with relational matrices and problematise the relationships of domination and exploitation based on the hierarchical classification across species, genders, ethnicities, class, ages, abilities and others. DT, and to some extent FPE, seek to make audible silenced voices and recognise marginalised people and beings in the production, administration and distribution of what the Western matrix of knowledge considers "resources". In this process, both approaches assert the notion of care and its relational dimension, seeking to transcend the dichotomous conception that separates the care of humans from the care of non-human others.

We have chosen to share our ideas under the format of a conversation, not only to engage the reader easily, but also because of our epistemic position. We position ourselves through our belief that knowledge is produced through dialogue—among authors, perspectives, currents of thought and cultural traditions over time, rather than through assertions which are refuted in a series of rebuttals.

The format of conversation allows us to present our perspectives and ideas in motion. We wrote the chapter on-line, over time and across different places, writing today from Argentina and The Netherlands. In this *reencuentro*, we revisit the dialogues and exchanges we have been having since 2014, when we worked together at the National Research Council in Argentina, enquiring into the idea of knowledge dialogues and prompting critical reflections on the dominant ways of producing knowledge. In this conversation we intertwine old with new stories and experiences, to advance our critical and reflective journey.

Although our ideas converge, there are also differences due to our personal way of expressing ideas and thinking about the dynamics we are discussing. We celebrate the polyphony of voices as part of our epistemic approach and of our collective practices (such as the writing of this chapter), trusting that our differences and singularities in dialogue complement each other and enrich what we want to share with the reader.

Let's Begin the Conversation…

Mariana: I would like to start this dialogue by revising the notion of urban space, the meanings and senses associated with it, and the way in which these are pragmatically constituted in the colonial strategy. In other words, we know that the idea of metropolis is central to the modern perspective, colliding with the ways of establishing community in other-than-western cultures such as Indigenous societies.

Ideas such as order, cleanliness, efficiency, ostentation, and property are manifested in a particular aesthetic in "urban environments". These ideas can conflict with practices and meanings of Indigenous settlements. This confrontation generates a border zone (an area of cultural translation), a liminal area that is difficult to understand in all its complexity.

From your personal and research experiences, how would you propose to consider these situations of liminality?

Agustina: When we speak of the urbanisation of the territory, we need to reflect on the modern strategies for territorial control that the nascent modern state of Argentina carried out in the military campaigns which attempted to exterminate Andean Indigenous peoples and cultures. These strategies were formed around the establishment of urban centres in the euphemistically called desert which the oppressors understood as needing to be civilised by modern citizens rather than "hostile savages".

"Shame is the repeated feeling in each of the stories"

The creation of urban centres in the South Andean territories can be described as the modern/colonial settling of land and peoples to control the territory. It was the project to civilise the "wild". Not only were urban settlements established but also National Parks were established across the southern Andes to control territories and their resources. Using the rhetoric of modernity and environmental preservation, the foundation of urban spaces was a way to control Indigenous communities. The establishment of National Parks forced Indigenous communities to move to the city to survive on a subsistence level. Yet, at the same time the urban space, and its innermost entity, the house, was also a place of resistance and re-existence for Indigenous communities. Since the urban public space became the surveillance space and the place where interactions were allowed only under a particular form, the intimacy of the home was the fundamental place to keep Indigenous memory, language and beliefs alive.

Urbanity was the means to impose a way of life and a way of being. Such "codes of urbanity" were tools of control and constant surveillance of Indigenous people by the Argentinian state. In these imposed modern urban spaces, a racist, disqualifying, omnipotent, moralising gaze permeated Indigenous peoples' lives through a permanent distinction between 'good' citizens who could read the codes of urbanity and those 'uncivilised' people who were seen as lacking the knowledge to read the codes.

These hundred-year-old stories continue in the present. The city of San Martín de los Andes is one of the urban centres established in the Southern Andes; it is an officially self-proclaimed intercultural city. Over the last years, public debates and negotiations around the management of the land located on the urban margins that was given back to the Indigenous communities gained visibility. The central question of debate was around the level of autonomy that Indigenous territories should have. The question was whether and how territories legally given back to the Indigenous communities be surveyed and measured. For example, should public streets be opened, and night lighting installed?

The question of interculturality brought with it notions of the "lack and excess" in how people read the codes of urbanity. If a marginal area of a town lacked public lighting, street layout and urban planning, it lacked urbanity. And at the same time, a large part of the Indigenous community was against those urban interventions.

Consider this: many marginal urban territories were flooded by mud, and it was difficult to get there by car. Yet, people resisted the development of streets. Why? I asked myself that many times when I was doing my PhD research. One day I saw a tourist with a 4 × 4 truck speeding across the area. I realised that the design of the settlement was not for the Indigenous peoples, the Mapuches, but for the tourists. In communal Indigenous territories there were no streets, people walked; there was no street lighting, people walked with torches. Streets and lights were not for Mapuches, but for tourists that could then use roads to visit villages. The rhetoric of urbanity hid the control of that Indigenous peripheral space. Urban development, based on the needs of people outside the Mapuche community, provided the justification to intervene on those territories and to satisfy other-than-Mapuche demands and expectations.

I spent the winter of 2015 and 2016 visiting the Mapuche Community located in the borders of the city of San Martin de los Andes during my PhD research. In June 2015, I was invited to the *We Tripantu* (a Mapuche celebration for the beginning of a new cycle) in the *Paraje Trompul*—a place located on the northern edge of the city that is part of the territories that had been removed and recently returned to the Mapuche community. I didn't know how to get there. It was a very steep one-way mountain road. I asked if people there shared their cars, but they didn't, they all walked. I arrived first so I waited at the school gate in the night darkness. Suddenly, in the distance, I saw little lights appear—like fireflies in the darkness of that cold winter dawn. They were the flashlights of the people walking together to the school.

Walking to school was a moment for conversation and encounter. Teachers, as they walked, shared what they planned to teach. This shared walking did not mean a lack of car as I had assumed. By crossing the city, climbing mountains, chatting, listening, sensing the territories together with others, humans and other-than-humans, they were welcoming and enjoying being in relation otherwise.

During the winter of 2016, different women from the Community shared with me stories about their lives, their childhoods, and the lives of their elders inside the Mapuche Community. One of them told me that when she was young, she used to walk from the urban margins where she lived to the centre of the city where she went to school. Every morning she walked down the mountain with her sister and her father. They often arrived class late with mud on their white school pinafores or *guardapolvos*.[1] Her sister used to run away from school feeling ashamed to enter the classroom with her white pinafore covered in mud. But my interlocutor, in contrast to her sister, proudly displayed the visible stains to her classmates, saying how exciting it was to live in the community. Another woman, a Mapuche teacher, told me that one rainy day, after washing her white school pinafore, she dried it in the wood-burning kitchen. The day after, inside the classroom a student, covering up his nose and making obvious gestures of disgust said: 'It smells like Mapuche in here!' referring to the smell of smoke.

In Argentina the white school pinafore holds a series of specific meanings: the homogenization of citizenship, the erasure of social and cultural differences and the identification with a civilised (white) westernised citizen. It is a symbol of national union and equality, modesty, hygiene and neatness. Due to its homogenising, standardising and disciplining character it has been the emblem of codes of urbanity in school settings. What is the meaning of my interlocutor pointing to her white school pinafore stained with mud? What is the meaning of leaving a white pinafore stained not clean but smoky, instead of ironed, perfumed and starched?

These daily life stories point to liminality or "thin cracks" through which the cultural heritage of an other-than-Western way of being in relation sneaks in. They are as fine as the cracks through which light filters, preventing blackness, preventing complete erasure. Even when codes of urbanity prevail, in the most intimate space, in everyday life practices, those powerful remnants of other ways of being in relation, persist. There may be minimal, subtle, silent fragments that most of the time go unnoticed. Yet, for that very reason, they survive.

[1] The white school pinafore, *guardapolvo blanco* in Spanish, is the uniform used in the public primary schools in Argentina. Inspired by hygienist precepts, the white school pinafore looks like the knee-length overcoat worn by professionals in the medical field. It is mandatory, worn by teachers (mostly women) and students. It has been more than 100 years since the state recommended its use, considering it both a democratising and disciplining element, for social inclusion and for educational organisation and control. It is, to this day, a symbol par excellence of free public education in Argentina.

The stories I heard in the Mapuche Community speak of lack or excess in relation to the norm. Loud music, grotesque laughter, those who speak loudly, don't work hard, those who walk rather than drive cars, those who are always muddy and dirty. These situations of liminality are transgressive. They are gestures that break away and defy what the rules of urbanity dictate. They reveal the persistence of another way of being in relation. If the mud is life, where the strength of the ancestors dwells, then to connect with the mud is to interact with forces beyond modern-Western understanding of life, death, space and time. Walking can mean the misfortune of lacking a car, but it can also mean the desire and the opportunity to relate with people and nature, with being in territory otherwise.

Mariana: It is beautiful what you say, subtle and powerful. It seems that there has been a set of representations—strongly supported by visual images—which coalesce in what is seen as an "Argentine identity". The wearing of a white school pinafore is linked to what is the right colour skin; the enforcement of short hair (in the case of men) and long combed hair (for women); a certain "moderation" in the forms, gestures and the way of speaking. Such aesthetic attributes are part of modern ethical values of control, discipline, work and effort... indispensable requirements to be and appear "Argentine"; to be people worthy of settling into national territory, of considering themselves owners of these lands and deploying their power.

Now, it is curious that national territory, particularly that of Southern Argentina, belongs—to a great extent—to foreign owners.[2] Argentine citizens have found it difficult to problematize the colonial erasure of the presence, legitimacy and territorial rights of communities pre-existing the conquest. Indeed, it is only in recent decades that Argentinians have become aware of such denials. But the situation is even more complex in the present since the information regarding who is owning Argentina's resources is now openly debated in the media. Along with the concern about economic dependency on international organisations is the growing awareness that the Argentinian territories themselves are owned—by law—by billionaires from the global North, who are icons of wealth and exploitation of people and resources.

[2] According to the National Registry of Rural Lands (RNTR) 5.57% of the rural Argentine territory, that is 16.253.279 hectares, is in the hands of foreigners (half of the surface of Italy). Moreover, 40% of the Argentine territory (around 65 million hectares) is in 1200 landowners' hands (INFOJUS, 2015).

How do you understand these growing concerns around ownership and exploitation of territories? Is there a need for collective questioning of the concept of tenure and property?

Agustina: To continue with your representations of what makes the Argentine citizen, I would like to add that all of those representations are supported by public laws and policies. For example, the laws that promoted "education on Argentine heritage": these laws banned Indigenous languages in schools and imposed Spanish as the only language of instruction. The only history taught was one of national heroes, through celebrations and hymns, idealising one specific citizenship: that of the white, Western(ised), urban man (not even white Westernised woman).

These colonial strategies gave legitimacy to the submission and exclusion of difference, even if their attempts were not entirely successful. The imposition of one language, one flag, one official history, one religion, expressed the control of difference and installed (in law and custom) a model of society based on domination—what the Peruvian sociologist Anibal Quijano has named coloniality (Quijano, 1992).

The legitimacy of the subjugation of difference led to the public denial, the rejection and the shame of being Indigenous, and therefore, of not "being worthy" of establishing an urban community of citizens. Although many privately resisted this subjugation, shame is the feeling that is repeated in each of the stories I heard. The shame is not something of the past, it is also experienced by the people of my generation who, during childhood, learned from their parents to hide, deny and to be ashamed of their Indigenous identity, language, memory and beliefs. The current processes of Indigenous re-identification and "dignification" (as people name it within the community) of being Mapuche, is linked, precisely, with the rejection of shame. I point to this to understand why it has been difficult to problematize the question of the presence, the legitimacy and the territorial rights of the communities that pre-existed conquest of the land. In recent decades, with the recognition by the State of the pre-existence of Indigenous communities to modern state, silenced stories are emerging which reveal that the community life in these peripheral territories existed before the white Western settlers who, with extraordinary effort and sacrifice, settled, populated and civilised those hostile lands as official history stated.

Within historical and geopolitical processes of colonial domination and independence in Latin America, the South Andean region is a periphery in relation to urban centres of power. The South Andean territories were geographically disconnected from the urban centres founded along the Royal Road (or *Camino Real* in Spanish) of the colonial period, the road that during the colonial period linked the Port of Buenos Aires with the Alto Peru. These territories entered modernity as an empty space on which to expand productive frontiers; a land rich in resources that had been wasted, land favourable for development, a space without a past and with full possibilities for the future. The systematic policy of transferring public land to private hands through donation, sale or reward for services rendered to the Nation (such as the financing of military campaigns) produced large concentrations of land in few hands, not only foreign but also local. The complicity between local elites and international interests consolidated colonial relations after independence. Once again, an illustration of what Quijano called coloniality.

We have now an opportunity to question collectively the ideas of possession and property. It is a far-reaching debate. It is important to question the notion of individual property and recover the idea of community property. It seems to me that the first step would be to rethink the collective, and from there, open the way to debate ownership. Since the appropriation of these territories by the national states, the land has been classified by its use and by who owns it. The notion of ownership is so strong that even the Indigenous communities that have effectively managed to legally return to their ancestral territories are considered owners. But for Indigenous people, land is not a property. However, to receive territorial restitution they had to accept the idea of ownership. Through this process, they managed to add a new figure to the constitution: Communal property. The figure of "Community property" was the way Indigenous communities were able to meet the requirements, within the limits of the modern state and with its legal tools (within modernity, but on its margins).

It is important to review our history and rethink our future. In this Southern Andean space/time, a series of highly conflictive situations converge. On the one hand, we can no longer continue ignoring, postponing, or denying the existence of Indigenous peoples and their pre-existence in these territories—according to the Constitution, they are the legitimate owners. On the other hand, we cannot ignore that the access to these lands and their resources have been handed over[3] to foreign millionaires and multinational corporations who legally own them. Although we cannot undo this completely, we can stop its advance. Local movements that have spent decades of continual struggles have reached national and international visibility.[4] This moment of profound civilizational crisis is an opportunity to reflect on the process and undo the restraining of other ways of seeing territory.

Mariana: These are situations of translation; that is to say, instances in which to be able to understand or to establish a dialogue, something must be "lost", in meaning or sense.

On the one hand, these processes—which go far beyond the mere issues of language—sink into cultural distances that integrate different ways of feeling and thinking. If Indigenous communities constantly renounce their own meaning making and senses (their deepest ways of inhabiting the world) it is worth asking ourselves what we are losing or are willing to lose, from the Western senses, in pursuit of an effective intercultural encounter. I am referring to how we transcend the meanings behind the control of people, land and body at the institutional level, which appear in the different areas of state administration. I understand

[3] The plundering, handover and "foreign takeover" (*extranjerización*) of lands has been going on for more than a century in Argentina, as have the claims and struggles of indigenous peoples to recover them. From 1880, the Argentinian State applied a "systematic policy of transferring public lands to private hands through the donation, sale or reward for services to the Nation [the killing of indigenous populations]" (Bandieri, 2005). Donations were established by law to encourage colonisation (Minieri, 2006). After the indigenous erasure at the end of the nineteenth century, the lands taken were ceded to investors, giving rise to the large livestock companies of foreign capital. This process took place within the framework of legal disorder that protected speculation and hoarding in the hands of foreign actors (Vazquez & Sili, 2017). The English company "The Argentine Southern Land Co." (renamed *Compañía de Tierras del Sud Argentino* at the time of the Malvinas war), for example, received almost a million hectares in Patagonia as a gift from the nation. By the end of the twentieth century, a new cycle of acquisitions by external investors was consolidated. The most emblematic example is the Benetton group that acquired the *Compañía de Tierras Sud Argentino*, accumulating 900,000 hectares (Forty times to the City of Buenos Aires.) (Minieri, 2006, pp. 7–9).

[4] In recent decades, denunciation of land foreignization has achieved national and international visibility and circulates in social discursivity. An example of this is the public support in 2004 of Adolfo Pérez Esquivel, who was awarded the Nobel Peace Prize in 1980, to the Mapuche people in their struggle to recover their ancestral lands, who stated that "If we don't stop this intrusion we will live in exile in our own land" (Macarenhas, 21 August 2006).

that this is a key question for those within social studies and humanities who are committed to diversity.

On the other hand, the distance between cultures does not come just from different meanings and senses but fundamentally from the way they are produced; something that goes far beyond language; but what we know as forms of "representation" of culture and meaning are the last thing the West would be willing to leave behind.

In academic fields, but also in the general social arena, the dominant representational modality is the verbal one, with all the semantic restrictions that this implies. It is enough to remember that scientific texts—rationally used, unlike in the field of arts, such as poetry or literature—are defined through a limited and limiting series of logical articulations. The format of academic texts requires a way of structuring that constrains the emergence of other (what are perceived as non-rational) senses. Indeed, those who decide to break these semiotic patterns of construction place themselves outside scientific knowledge, appearing as art or in a diverse cultural framework, with all the institutional exclusion that this implies.

The Andean cultural horizon is simply not comprehensible by Western rationality. It cannot be covered from the logical and linear linguistic sign because the deep matrices of its senses are different, strongly marked by the notion of circularity. It is a founding pattern that appears both in its landscapes and in the way verbal discourse is displayed. It is manifested in its iconography (which entirely seals the mountain range by petroglyphs), ceramics and the idea of inhabiting itself, which, from a Western perspective, would be linked to architecture, urban planning and conservation. Only through ethical and aesthetical assessments of multiple materialities would it be possible to take a glance at Andean epistemology from these otherwise forms of representation (pictorial, sculptural and architectural).

Given this, how do you understand the different ways of inhabiting, between the Indigenous cultures of South Andes and the Western cultural matrix?

Agustina: You ask what we are willing to lose from the Western knowledge to allow a true intercultural encounter. Your question inspires me to ask a further question—how much more will we lose if we do not consider other perspectives? The Mapuche poet Liliana Ancalao (2018), wrote:

> I'm talking about an ancient language and the ignorance of men who mapped a country over a territory full of names, elements, and meanings, silencing it. I'm talking about what we lost. All of us.

> All of us who were born without knowing the names of every plant, every stone, and every bird of this land.
>
> I woke up in the middle of a lake. In gasps I tried to give thanks but didn't know the words.

I resonate with what Ancalao raises about the plurality that has been deprived to "all of us"—the knowledge that all of us lost. What would we all gain with an effective intercultural encounter?

Looking at the different ways of inhabiting helps us to reflect on what you expressed about the distance between cultures as a consequence of the way of producing meanings and senses; and not as a result of the different meanings and senses themselves. Under a broad conception, inhabiting the world, as a practice, is a continuous process anchored to a body, a territory and a specific social and historical context. Inhabiting the world is a collective process embedded in a network of relationships woven through time and marked by a particular way of relating. In this sense, based on my own journey, I understand that the deep network of relationships between the territorial, the ancestral, the spiritual and the communal, is central to understanding the inhabiting of the South Andean Indigenous cultures.

Following up on your comment around circularity as a relational matrix, I can add that the gap between the different ways of being in the world—in this case between South Andean Indigenous culture and Western culture—reflects the conflicting distance between the founding elements of their relational matrices.

Without intending to essentialize, and without ignoring the complexities and heterogeneities within them, it is understood that the Andean relational matrix is based on a life-affirming relationship that is sustained in the care of all beings and the care for the reproduction of life. Contrary, and given its colonial origin, the relationship of domination and control of nature and people for exploitation and consumption are constitutive of the modern Western relational matrix. This last clearly does not represent all of Western culture. There are deep life-affirming Western struggles that question and manifest against this relational mode.

The Andean relational matrix sustained in care is based on the conception of equality between the various forms of life and therefore on the intrinsic right of the different forms of life to regenerate themselves. In contrast, the most predominant relational matrix of the West, sustained in domination, is based on the conception of hierarchies between humans (since the invention of the idea of race), and between humans and nature. The Andean world view is based on a circular perspective in which human beings are part of the weave, while the Western is grounded in a linear hierarchical perspective in which "The Man" (male, Western, urban, white, heterosexual, able bodied) takes the highest place.

Care for life and its reproduction contrasts with violence produced by domination and exploitation—founding characteristics of a relational matrix that since the emergence of the Enlightenment and coloniality dominated the world. This implies moving from the centre the care for the human and for the individual, to bring instead the care for the capacity reproduction of the complete weave of life. Care for the plurality of life seeks to protect the complete weave of life. Ontologically, each being in this weave occupies a vital place for continuity. Therefore, to protect life is to make room, to keep a safe space for each being to develop its life course and to regenerate. Circularity, from the Andean perspective, is going back to the origin. This stands in contrast to the Western Enlightenment conception of unlimited growth in pursuit of progress and (economic and political) development, in which the direction is linear, away from the origins. Posing the idea of returning to the origin helps us to understand how growth can come from the cultivation of life. Learning from the experience of the South Andean cultural horizon invites us to make room for the plurality of worlds and to rethink modern/colonial Western practices. This implies positioning the West as a place of reception, letting be rather than doing, becoming affected and therefore becoming vulnerable, letting vital processes take their course as decolonial and feminist political ecology approaches invite us to do.

Mariana: Isn't it remarkable the way in which these considerations have reappeared, although we are not sure precisely how or where from, and have now begun to circulate in general social discourse? Discussions on (the promotion of) interculturality seem to have generated the possibilities of emergence of this kind of engaged understanding with the otherwise to the West. I would caution us here, there is a trap. The notion of "worldview" has served to position non-Western cultural heritages such as the Andean worldview in a space of distant gnoseological recognition, which means it cannot question the legitimacy of dominant Western institutions, governed by their epistemological rules, even those which are critical.

For instance, each of the axiological, ontological and semiotic concerns you raise necessarily become an epistemic model—a dynamic of knowledge that starts from a way of perceiving which appears completely unknown.

Take the question of naming the "natural world": from the Western epistemological matrix of knowledge and its way of representing, plants are only considered in their ornamental dimension or in their instrumental possibilities. Biology, for example, has made—and continues to make—enormous efforts to classify minutely the South American flora to find its medicinal properties. Thus, each of the "discovered" varieties are observed and categorised according to these specific kinds of qualities: emmenagogues, healing, anti-fever, etc.

It is difficult to find a clearer example of the instrumental perception of the world, so lucidly denounced by critical philosophy since the late twentieth century. However, this pattern of knowledge remains completely intact in a field that is as important as medicine and related disciplines, such as biology or physics.

From the Andean worldview, this cognitive operation and way of approaching the natural world, implies nothing more and nothing less than losing the opportunity to encounter the different processes and experiences that each plant displays, of which those related to the physical, biochemical dimension, are just a small part.

It is important to be clear that cultural difference in the way of perceiving and understanding the world of "nature" is perfectly valid from the different perspectives through which the body is understood. But what is important to underline is that in an Andean perspective, the body (one's own and others) is nothing more than a point in that knowing weave.

So, the difference can seem immeasurable, impossible to overcome; but this key epistemological feature of Andean culture—the circular knowing weave—is not given its epistemic status by the West but is referred to as part of an almost picturesque and distant "worldview".

It is also intriguing how the body—that great unknown—has been returning in different ways to the scene of social theory, demanding its denied place. Anthropology of the body and emotions has risen as a subject of study in Performance Studies in dialogue with feminist theories.

How do you consider the place that the body has reached today in social studies and what elements do you think it is possible to contribute from the Andean cultural horizon?

Agustina: Your thoughts about plants recall for me a question: what is the purpose of building bridges across the plurality of knowledges? What is the purpose of the knowledge that we produce by encountering others? I continually return to these questions, trying to be cautious with the knowledge that others shared with me and with what I do with it from my privileged position in the academy.

As a decolonial researcher, I orient my academic efforts towards encountering perspectives otherwise to learn from them. However, this learning is not aimed at recovering specific knowledge about something in order to possess it or obtain something from it. Instead, I aim to learn about other ways of being in relation.

Mapuche women use plants for healing purposes, but they also relate to them in a very particular way. Before taking a plant from the ground, they ask permission to use it and benefit from its properties. Both the ritual of asking for permission, of expressing in words the reasons why the action is carried out, and the ritual of thanking the plant for providing its properties, have to do with the relational matrix between women and healing plants. The ritual accounts for the consideration of plants as entities, as a living part of the weave of life with rights to be and reproduce themselves. As Quijano says "it is not accidental that knowledge was considered then in the same way as property — as a relation between one individual and something else" (Quijano, 2007, p. 173). In this ritual example, the idea of the subject of knowledge and object of knowledge (and therefore objects of domination and exploitation) are diluted. The plants become subjects and the relationship is one of cooperation between subjects.

But, returning to your question, I understand that the body, like the territory, plays a preponderant role in knowledge production. When understanding the notion of inhabiting the world, as a process sustained in a historical, cultural, social and political weave that shapes and transforms it; the notions of body and territory become central to the deep intertwined meanings inscribed in dwelling. Knowledge is always situated in a territory and in a particular body that produces it. By accounting for the geopolitical and body-political location of the subject who speaks, we recover the partiality of perspective and in so doing can question the universality of knowledge. The situated character of knowledge (Castro-Gómez, 2005; Haraway, 1988) and the focus on the violence of Western domination inflicted on other than heterosexual, white, male bodies are the main points of convergence between feminism and the decolonial perspective.

From the Andean cultural horizon, it is possible to problematize the ownership of the body and reflect that, like the territory, the body is not a property. And, in the logics of coloniality, like the territories, racialized bodies have been property.

The Andean horizon shows us another way of understanding the body and territory, both part of a common weave. Andean cosmology teaches us to see that it is not possible to separate the body/territory from the spirit, and that each living being (body/spirit) has a place and fulfils a vital function in the weave of life. The Andean cultural horizon teaches us about our responsibility for care so that life can continue its regenerative process. From this perspective, care for the bodies/territories has a collective nature—it is caring for the indivisible weave of life.

Mariana: The expression you use regarding the Andean cultural horizon is very interesting: when you say that it "teaches us", it reflects it's the idea's deepest sense on two levels: on one hand, it disobeys that dangerous idea, inherited from early Anthropology, which postulates that it is possible to know other cultural frameworks without questioning our own semiotic and cognitive habits. This early anthropological assumption has been strongly problematized from the decolonial perspective (Segato, 2018). But on the other hand, the idea of "letting ourselves be taught by other cultures" places science in a completely different position. It is no more about a science that "knows" and goes in search of its multiple "objects" or "subjects" of study to consequently "teach" the rest of society; on the contrary, it begins from an intentionally receptive position—less masculine and less imposing—to put Western science on an equal footing, from a dialogical position, with respect to what it is trying to address.

> ***"to hold differences, be taught"***

As such, this move constitutes an important feature of the decolonial movement as different from other theories which are critical of modernity; it aims to recover the cultural heritages silenced by the colonial enterprise.

Further, this emancipatory decolonial movement is not restricted to scientific production. The issue of care that you mentioned before is intimately related to the concepts of reciprocity and cooperation, which go completely against the grain of the current social scientific dynamics that are open only to a certain type of production of logical argumentation.

Research "products" (papers, thesis reports, books, etc.) are not materials or media designed to communicate with diverse cultural groups. Quite the contrary, these pieces strengthen intra-institutional communication, fuelled by information that comes from 'international' knowledge, perpetuating the extractive-academic-logic.

How do you understand decolonial research as an opportunity to "let oneself be taught" when the deep structures, the hierarchical dynamics—semiotic and socio-cultural—that we intend to transform by the decolonial move have not yet been modified?

Agustina: The desire to "let ourselves be taught" is profoundly valuable, although still unattainable. Bridging the distance that exists between those desires and the possibilities of learning from other relational matrices is where we are lacking. Reducing the expectations of what Western epistemology can do would be the first step in this direction.

The profound ancestral wisdom that inhabits the Andean cultural horizon is there. Yet, the very idea of "ancestral wisdom" refers to an almost mystical idea that the West find captivating and attractive, but continues to decontextualize, simplify and abstract from its epistemic status, through the notion of "worldview", as you explain above. Moving towards an encounter with the deep and unintelligible processes that the southern Andean ancestral wisdom holds means establishing a dialogue, which is not easy.

"Research outputs" are specific products resulting from research processes. Even in the research carried out by those positioned on the margins of modern Western epistemology, investigative "products" continue to be reductions, abstractions, translations into an academic language, of an entire universe that has been opened and shared to us. When translating lived experiences, we reduce all that richness to what (from our own perspective) we can understand. All the abundance that comes from other processes, all that immensity that exceeds the scope of our understanding frameworks, becomes negligible or unintelligible, or "in-significant" as Zulma Palermo (2004) describes it.

The desire to build bridges with other ways of being and relating pursues a profoundly valuable goal: epistemic, social and environmental justice. However, in the movement towards understanding, interpreting, and translating those interpretations to a specific language, all that shared richness is reduced and simplified. This universe of meanings alludes to something much deeper than we (the Western-based scholar) think we encompass. Can we understand from our Western interpretative matrix the deep relational processes that unfold in the ritual of asking permission and thanking the plants before extracting their medicine from them? Even if we cannot fully understand it, we should not reject the invitation to let ourselves be taught.

These reflections seek to contribute to bridging those gaps between diverse cosmologies. My encounter with others made possible my awareness of coloniality. My encounter with other perspectives, in specific territories and with specific bodies that have suffered (and continue to suffer) in their own flesh the colonial wounds, has opened up the possibility of becoming aware of processes that were foreign to my Western experience; to unveil previously naturalised situations. Encountering others transforms one's perspective. After knowing the experiences marked by coloniality it is not possible to be indifferent to it.

Even when the deep structures have not been modified, the invitation to let oneself "be taught", and to then return to academia to reflect epistemologically on what has been learned together with those bodies that have been traversed by coloniality, has transformative potential. This practice is within the framework of our possibilities and contributes to the emergence of the plurality of the world. The path we move on is that of the fight against oblivion, based on the recovery and reconstitution of the memory that inhabits plural bodies and territories. We could move forward by making a turn, from an active position to a passive one, by listening, by letting oneself be permeated by the experiences of others. The challenge would then be to lean on the memories and stories that others share with us, nourish ourselves with them, and work hard to keep them active, alive; against the destruction of the plural heritage and the extraction of life at all levels, human and other-than-humans.

Among our effective possibilities of generating transformations is that of enriching a critical movement that contributes to the struggle so that others can take part. This does not end simply in criticism. The critical review would be a reflection on ourselves. We would go to meet others, but this time not to understand what those others think and establish intercultural and inter-gnoseological dialogues with them (something that seems ambitious and unrealizable), but to revisit what we think of our own trajectories once we articulate with others. And that would require expanding our reception and listening capacity. The invitation to let oneself "be taught" is the invitation to become available, permeable, sensitive—to make room to host differences.

A Few Closing Words Until We Meet Again…

Throughout this conversation we have exchanged considerations around some of the most challenging questions of our time—starting from the consideration of the "urban" in the Andean-colonial encounter, we moved into the question of knowledge in its different manifestations and ended with the need to rethink and listen otherwise.

Listening to the Andean perspective, which is not homogeneous but singular and characteristic of an entire "space-culture", we have tried to think with the readers about questions related to the different ways of inhabiting territory, understood from a broad and circular conception, as an extension of our own body.

The Andean perspective understands the individual and collective are part of the general body of nature within a circular matrix. This matrix associates the body with abstract and symbolic, mental or cultural processes through a bond of mutual affection.

We think that the investigation and understanding of these issues is currently emerging as particularly challenging, configuring a space of knowledge that is given voice in different spaces of research and discipline such as Psychoanalysis, Gender Studies or Decolonial Studies at times overlapping with FPE, by bringing centrality to the (collective or individual) body and sensory, by calling attention to care for humans and other-than-humans, and fundamentally, by striving towards being, doing and knowing otherwise. In this sense, this space of knowledge is open to knowing and wisdom coming from non-Western spaces, such as Andean Indigenous cultures, that as we have tried to show through this conversation, have essential contributions to this moment of re-founding a new way of understanding life.

References

Ancalao, L. (2018, January). El idioma silenciado. *World Literature Today*. https://www.worldliteraturetoday.org/2018/january/silenced-language-liliana-ancalao

Bandieri, S. (2005). Del discurso poblador a la praxis latifundista: la distribución de la tierra pública en la Patagonia. *Mundo Agrario, 6*(11). https://www.redalyc.org/articulo.oa?id=84501101

Castro-Gómez, S. (2005). *La Hybris del Punto Cero: ciencia, raza e Ilustración en la Nueva Granada (1750–1816)*. Editorial Pontificia Universidad Javeriana.

Escobar, A. (2007). Worlds and knowledges otherwise: The Latin American modernity/coloniality research program. *Cultural Studies, 21*(2–3), 179–210. https://doi.org/10.1080/09502380601162506

Haraway, D. (1988). Situated knowledges: The science question in feminism and the privilege of partial perspective. *Feminist Studies, 14*(3), 575–599.

INFOJUS. (2015). *Registro Nacional de Tierras Rurales. Una política registral para la soberanía territorial*. Editorial Ministerio de Justicia y Derechos Humanos de la Nación. http://www.saij.gob.ar/docs-f/ediciones/libros/Registro_Nacional_Tierras_Rurales.pdf

Macarenhas, T. (2006, August 1). *New Internationalist*. https://newint.org/columns/essays/2006/08/01/patagonia

Minieri, R. (2006). *Ese ajeno sur*. Fondo Editorial Rionegrina. https://ia800803.us.archive.org/24/items/RamnMinieriEseAjenoSur/libro%20minieri%20ese%20ajeno%20sur.pdf

Palermo, Z. (2004). Ricardo J. Kaliman, Alhajita es tu canto. El capital simbólico de Atahualpa Yupanqui. *Revista de crítica literaria latinoamericana, 30*(60), 392–394.

Palermo, Z., & Quintero, P. (2014). *Aníbal Quijano: textos de fundación*. Ediciones del Signo.

Quijano, A. (1992). Colonialidad y Modernidad/Racionalidad. *Perú Indígena, 13*(29), 11–20.

Quijano, A. (2007). Coloniality and Modernity/Rationality. *Cultural Studies, 21*(2–3), 168–178. https://doi.org/10.1080/09502380601164353

Segato, R. (2018). Una paradoja del relativismo: El discurso racional de la antropología frente a lo sagrado. In F. Gorbach & M. Rufer (Eds.), *(In)disciplinar la investigación: El archivo, el campo, investigación, escritura y producción de evidencia* (pp. 25–63). Editorial Autónoma Metropolitana.

Vazquez, A., & Sili, M. (2017). Dinámica espacial del proceso de extranjerización de la tierra en la Patagonia. *Journal of Latin American Geography, 16*(2), 117–137.

Index

W. Harcourt et al. (eds.), *Contours of Feminist Political Ecology*, Gender, Development and Social Change,
https://doi.org/10.1007/978-3-031-20928-4

FSC
www.fsc.org
MIX
Papier | Fördert
gute Waldnutzung
FSC® C083411

Kolibri 360 GmbH
Ferdinand-Jühlke-Straße 7
99095 Erfurt, Deutschland
produktsicherheit@kolibri360.de